FACING GLOBAL DIGITAL REVOLUTION

PROCEEDINGS OF THE 1ST INTERNATIONAL CONFERENCE ON ECONOMICS, MANAGEMENT, AND ACCOUNTING (BES 2019), JULY 10, 2019, SEMARANG, INDONESIA

Facing Global Digital Revolution

Editors

Dyah Nirmala Arum Janie, Hendrati Dwi Mulyaningsih & Ani Wahyu Rachmawati

LONDON AND NEW YORK

Routledge is an imprint of the Taylor & Francis Group, an informa business

Typeset by Integra Software Services Pvt. Ltd., Pondicherry, India

Publisher's Note
The publisher has gone to great lengths to ensure the quality of this reprint but points out that some imperfections in the original copies may be apparent.

Library of Congress Cataloging-in-Publication Data
Names: International Conference on Economics, Management, and Accounting (1st : 2019 : Semarang, Indonesia) | Nirmala Arum Janie, Dyah, editor. | Dwi Mulyaningsih, Hendrati, editor. | Wahyu Rachmawati, Ani, editor.
Title: Facing global digital revolution / editors, Dyah Nirmala Arum Janie, Hendrati Dwi Mulyaningsih & Ani Wahyu Rachmawati.
Description: Boca Raton : CRC Press, Taylor & Francis Group, [2020] | "Proceedings of the 1st International Conference on Economics, Management, and Accounting (BES 2019), July 10, 2019, Semarang, Indonesia"--Title page. | Includes bibliographical references and index.
Identifiers: LCCN 2020005758 (print) | LCCN 2020005759 (ebook) | ISBN 9780367339128 (hardback) | ISBN 9780429322808 (ebook)
Subjects: LCSH: Economics--Congresses. | Strategic planning--Congresses. | Accounting--Congresses.
Classification: LCC HB21 .I618 2019 (print) | LCC HB21 (ebook) | DDC 330--dc23
LC record available at https://lccn.loc.gov/2020005758
LC ebook record available at https://lccn.loc.gov/2020005759

Published by: CRC Press/Balkema
Schipholweg 107C, 2316XC Leiden, The Netherlands
e-mail: Pub.NL@taylorandfrancis.com
www.crcpress.com – www.taylorandfrancis.com

First issued in paperback 2021

ISBN 13: 978-1-03-224062-6 (pbk)
ISBN 13: 978-0-367-33912-8 (hbk)

DOI: https://doi.org/10.1201/9780429322808

Facing Global Digital Revolution – Nirmala Arum Janie, Dwi Mulyaningsih & Wahyu Rachmawati (eds)
© 2020 Taylor & Francis Group, London, ISBN 978-0-367-33912-8

Table of contents

Facing Global Digital Revolution – Nirmala Arum Janie,
Dwi Mulyaningsih & Wahyu Rachmawati (eds)
© 2020 Taylor & Francis Group, London, ISBN 978-0-367-33912-8

Foreword

In terms of industrial revolution 4.0, what technological aspect can disrupt jobs nowadays and change the business model? The digital revolution has re-imagined how individuals live. Technology is changing the traditional business model and reinterpreting the value for the customer in economic circumstances. Kavadia et al. (2016) contradict the idea that technology is exclusively responsible for the transformation of the industry. Westerman (2017) supports that view, noticing that "technology doesn't provide value to a business, but that technology's value comes from doing business differently because technology makes it possible."

Facing the digital revolution has shown how business, economy, and management have been affected by technological changes in a more personalized product/service offering, a closed-loop process, asset sharing, usage-based pricing, and a more collaborative ecosystem, an agile and adaptive organization.

The 1st Business and Economic Series: Economics, Management, and Accounting Conference 2019 (1st BES 2019) presented various articles that discuss research in business, economy, management, and accounting in a digital era. The research needs to be emphasized by a continuous study for a more significant impact.

Dyah Nirmala Arum Janie
Hendrati Dwi Mulyaningsih
Ani Wahyu Rachmawati

Facing Global Digital Revolution – Nirmala Arum Janie,
Dwi Mulyaningsih & Wahyu Rachmawati (eds)
© 2020 Taylor & Francis Group, London, ISBN 978-0-367-33912-8

Organizing Committee

General Chair
Dyah Nirmala Arum Janie

General Co-Chairs
Aria Hendrawan
Anna Dian Savitri
Subaidah Ratna Juita
Tatas Transinata
Hendrati Dwi Mulyaningsih

Conference Coordinator
Santi Rahmawati
Ani Rachmawati
Febrialdy Hendratawan

Conference Support
Almas Nabili Imanina

Information and Technology Support by Scholarvein Team

Facing Global Digital Revolution – Nirmala Arum Janie,
Dwi Mulyaningsih & Wahyu Rachmawati (eds)
© 2020 Taylor & Francis Group, London, ISBN 978-0-367-33912-8

Scientific Review Committee

J.Vignesh Kumar
National Institute for Research in Tuberculosis (NIRT), Indian Council of Medical Research (ICMR), India

Hardani Widhiastuti
Universitas Semarang

Jerny Dase
Hasanuddin University, Indonesia

Kim Alvin C. De Lara
Department of Education- Division of Rizal, Philippines

Amri Panahatan Sihotang
Universitas Semarang

Rohadi
Universitas Semarang

Bhargavi Kaveti
Kakatiya University, Telangana India

Liberty Nyete
University of Venda, South Africa

Tay Kok Wai
Universiti Kebangsaan Malaysia, Malaysia

Sruthi V.S.
Jawaharlal Nehru University, India

Garry Kuan Pei Ern
Universiti Sains Malaysia, Malaysia

Endah Pujiastuti
Universitas Semarang

Facing Global Digital Revolution – Nirmala Arum Janie, Dwi Mulyaningsih & Wahyu Rachmawati (eds)
 ISBN 978-0-367-33912-8

Profiling the work force: Spearheading human capital management in the era of Industry 4.0

Ahmad Rozelan Yunus
Universiti Teknikal, Melaka, Malaysia

ABSTRACT: Industry 4.0 (IR 4.0) is the implementation of Cyber Physical Systems smart manufacturing for industrial production. Its impact in the manufacturing sector and other technology sectors is well documented. However, the impact to the human resource sector is understudied. Managing the human capital of IR 4.0 is not an easy task since it requires continuous innovation and learning dependent on people and enterprise's capabilities. Personality profiling strategy for appropriate worker screening approaches can play a vital role in the development of dynamic capabilities in an organization. This paper aims to present worker screening approach using systematic psychometrics profiling that promotes a climate of innovation and learning in organizations, and hence facilitates businesses matching the pace of IR 4.0. The overview employs the application of Holland Individual–Environmental Congruency career interest concept and adapts various psychometrics profiling approaches in proposing an integrated personality framework as an initiative towards an adequate systematic screening strategy for recruitment, placement, succession planning, and planning organizations' personnel development training for IR 4.0.

Keywords: personality profiling, human resource management, IR 4.0

1 INTRODUCTION

Personality profiling is a systematic process to record and analyze employees' personality traits. By understanding individual personalities, the organizational management team can better understand what influences personal and social life behaviors. Incompatible traits and characteristics can effect performance (Holland, 1997).

Industry 4.0 (IR 4.0) is the implementation of Cyber Physical Systems smart manufacturing for industrial production. Managing human capital in the era of IR 4.0 is not an easy task since it requires continuous innovation and learning dependent on the capabilities of people and enterprises.

Personality profiling for screening the right workers can become an important strategic approach in the development of dynamics capabilities in an organization. This article aims to present worker screening approaches using systematic psychometrics profiling that promotes a climate of innovation and learning in organizations, and hence facilitates businesses matching the pace of IR 4.0. Profiling is important in giving an overview to the organization's top management for improving and developing modules that are necessary for future individual development programs such as training modules, intervention programs, performance enhancement programs, organizational succession planning, career development programs (Yunus, 2004), as well as enhancing work performance and job promotion exercises. Thus, this work promotes a platform to profile individuals for the work place. Using Holland's (1958; 1997) Individual–Environmental Congruence Theory as the basis for matching individuals to their work environment has become the main fundamental concept of the proposed profiling tool.

2 INDUSTRY 4.0 ORGANIZATION SCENARIO

In an organizational environment, it is important to profile the work force (Yusoff, Bakarb, & Alias, 2006), especially in this era of Industry 4.0. Evidences in the extant literature have shown that personality characteristics play an important role in an individual's work performance and achievement. Unfortunately, the literature deals mainly with individual personality characteristics. Thus the ability to provide holistic and integrated personality domains is needed in dynamic and changing organizations (Yusof et al., 2016).

Therefore, an effort should be made to develop holistic and integrated personality profiling tools as a mechanism for properly screening the work force. Failing to recognize the right worker for the job will end up with recruiting and placing an inappropriate and incompetent worker into the job. A personality type would only be considered undesirable to the extent to which it counters the work performance expectations of a particular work environment (Holland, 1999).

3 INDIVIDUAL–ENVIRONMENTAL CONGRUENCE THEORY

The relationship between personality characteristics and work performance are well established in the literature. John Holland's Individual–Environmental Congruence Theory is regarded as the most influential in the field of career counseling (Brown, 2002). Therefore, Holland's theory and the subsequent research on it were explored to determine an appropriate means of understanding the behavior of organization members.

Holland concentrated on the differences between individuals, rather than their similarities. He defines six types of individual (realistic, investigative, artistic, social, enterprising, and conventional) and recognizes that these types will have different occupational interests.

Since its emergence more than fifty years ago, Holland's theory has become a major force in applied psychology. It emphasized the "searching" aspects of person–environment fit: "The person making a vocational choice in a sense searches for situations which satisfy his hierarchy of adjustive orientations" (Holland, 1997). There was also an emphasis on the acquisition and processing of environmental information. "Persons with more information about occupational environments make more adequate choices than do persons with less information." A precursor article on the Vocational Preference Inventory (VPI) in Holland (1958) describes the core of the theory - the projection of one's personality onto the world of work. The choice of an occupation is an expressive act that reflects the person's motivation, knowledge, personality, and ability. Occupations represent a way of life, an environment rather than a set of isolated work functions or skills. For example, to work as a carpenter means to have a certain status, community role, and a special pattern of living. This Individual–Environmental Congruency Theory has become the main fundamental concept of proposed profiling tool.

4 THE TOOL: i-PRO

i-PRO is Integrated Personality Profiling System. The concept of the tool proposed embedded various domains as shown in Figure 1 below. It covers a combination of personality traits domains and the working environment domains.

Domains of the the profiling developed covers four domains. Three main domains are 1) personality; 2) competency; and 3) core values, and "ethics & spiritual" serves as domain across. The three main domains break into sub-domains as shown in Table 1 below.

These profiling domains were adapted from various instruments such as Holland Person-Environment Fit, Hogan Personality, and Myer-Briggs Type Indicator. The Big Five Personality was adapted as the main base theory. It also embedded the organization core values, the institution's twenty-year strategic plan, and the Malaysian Education Blueprint known as Pelan Pembangunan Pendidikan Malaysia, Pendidikan Tinggi (PPPM PT). All these values

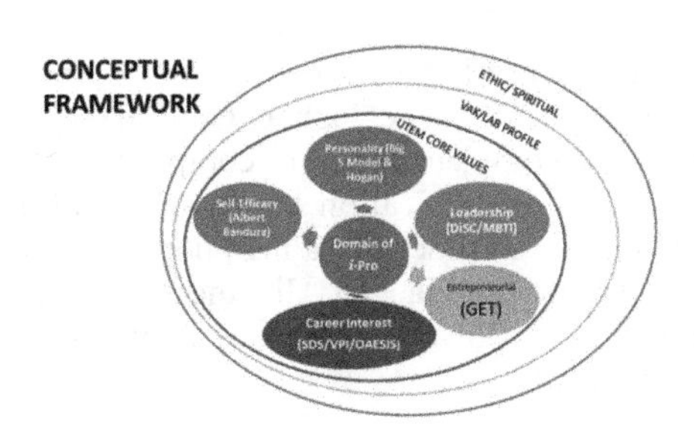

Figure 1. i-PRO personality conceptual framework.

Table 1. *i*-PRO main domains and sub-domains.

Main domains	Personality	Competency	Core values
SUB DOMAINS	1. Self Orentation 2. Career Orientation 3. Psycho-social 4. Dominant brain 5. Self Representation 6. Leadership	1. Entrepreneurial 2. Interpersonal 3. National Identity	1. Readiness for change 2. Resilience 3. Loyalty 4. Integrity 5. Professionalism
DOMAIN ACROSS	Ethics & Spiritual		

have been integrated as a fundamental domain of an instrument tailored to needs of the organization (Musa et al., 2016). Thus, this research aimed to develop and validate items in the instrument for the people/citizens in an organization.

4.1 *Person–environment congruency*

Applying the person–environment fit concept into the developed tools are elaborated as follows. The mean score for every personality type based on area of specialization in a studied organization were obtained by using descriptive methods. By using the three highest mean scores, Holland's three-code typology for specific departments from the organization has been determined. Holland's three-code typology (Holland, 1997) was also determined for the department itself. These codes presented the personality profiles of the workers for their working environment. Based on Table 2, the three types of personality with the highest mean scores for the studied organization were I (4.42), S (4.16) and R (4.02). This means that, the Holland's three-code typology for this organization, working environment is ISR.

4.2 *Congruency between personality types and work environment*

To determine the congruency level of individuals to the studied organization, The Iachan Congruency Index (1984) was used to calculate degree of congruency by comparing the code

Table 2. Workers' personality profile.

R	I	A	S	E	C	Holland's three-code Typology
4.02	4.42	2.23	4.16	3.08	2.81	ISR

obtained with the three-letter Holland letter from the score scoring the Opportunities Finder (Rosen, Holmberg, & Holland, 1994). The Iachan Congruency Index (1984) became the researcher's choice because it is a complete degree of compatibility and is suitable for measuring two and three-letter codes in the classification system (Holland, 1997).

In addition, the Iachan Congruency Index has a high degree of correlation index of r = 0.74 (Miller & Mark, 1992). Furthermore, the Iachan Alignment Index is an accurate measurement of the degree to be used in the field of research. An example of setting congruency degree using the Iachan Congruency Index:

Respondent Code	- R I E
Environment Code by Holland	- R S I

Based on Table 3, the respondents score "R, I and E", and the environment code score according to Holland is "R, S and I".

Based on Figure 1, the respondent code is arranged horizontally from left to right, and the program scores according to Holland are arranged vertically from top to bottom. Code "R" ranks as the primary code, rated 22 points, and code "I" rates 2 points (i.e., as a secondary code for the respondent code, and the tertiary code for program code according to Holland). The Iachan congruency level score of this respondent is 24 (22 + 2 = 24).

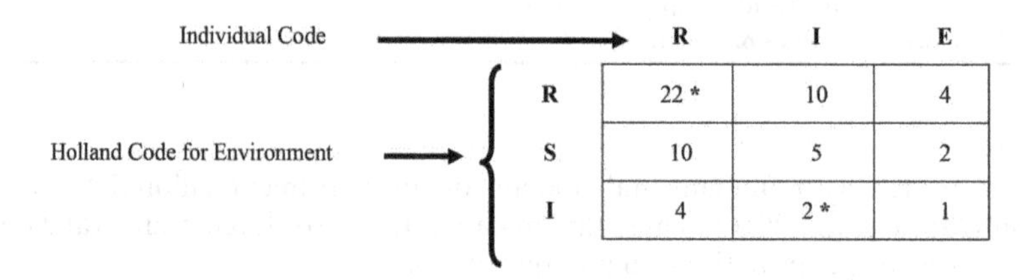

Figure 2. Holland congruency and Iachan congruency index level.
Source: Naemah (2007)

5 CONCLUSION

This systematic psychometrics profiling would be useful in promoting a climate of innovation and learning in organizations, and hence be able to facilitate businesses in the era IR 4.0. It employs the application of the Holland Individual–Environmental Congruency career interest concept and adapts various psychometrics profiling approaches in proposing an integrated personality framework as an initiative toward adequate systematic screening strategy for recruitment, placement, succession planning, and planning the organizations' personnel development training for IR 4.0.

ACKNOWLEDGMENT

The author wishes to express gratitude to Universiti Teknikal Malaysia Melaka (UTeM) for its support in getting the data and material in the advancement of this work and will likewise need to thank the unknown contributors whose remarks prompted an enhanced presentation of this work. In conclusion I additionally thank to UTeM for supporting this work under The Short-Term Grant Research Scheme PJP/2017/IPTK/S01521.

REFERENCES

Brown, D. (2002) *Career Choice and Development*. John Wiley & Sons.

Holland, J. L. (1958) "A personality inventory employing occupational titles." *Journal of Applied Psychology*. American Psychological Association, 42(5), p. 336.

Holland, J. L. (1997) *Making Vocational Choices: A Theory of Vocational Personalities and Work Environments*. Psychological Assessment Resources.

Holland, J. L. (1999) "Why interest inventories are also personality inventories." Davies-Black Publishing.

Iachan, R. (1984) "A measure of agreement for use with the Holland classification system." *Journal of Vocational Behavior*. Elsevier, 24(2), pp. 133–141.

Musa, H. *et al.* (2016) "The characteristics of users in the adoption of low loss microwave transmission glass: A conceptual paper." *Procedia-Social and Behavioral Sciences*. Elsevier, 219, pp. 548–554.

Rosen, D., Holmberg, K. and Holland, J. L. (1994) *The Educational Opportunities Finder: For use with the Self-Directed Search and the Vocational Preference Inventory*. Psychological Assessment Resources.

Yunus, A. R. (2004) "Personality congruence and compliance: A study among students of a higher learning institution in Sabah." Unpublished Fundamental Research Report. Universiti Malaysia Sabah, Kota Kinabalu.

Yusof, N. *et al.* (2016) "Relationship between emotional intelligence and university students' attitude." *Journal of Social Sciences & Humanities*, 24, pp. 119–130.

Yusoff, H. B. H., Bakarb, A. Z. A., and Alias, R. A. (2006) "Polygraphic counterproductive behavior index profiling system." in *Proceedings of the Postgraduate Annual Research Seminar*, p. 308.

Facing Global Digital Revolution – Nirmala Arum Janie, Dwi Mulyaningsih & Wahyu Rachmawati (eds)
© 2020 Taylor & Francis Group, London, ISBN 978-0-367-33912-8

Innovation strategy: How to answer the challenges of the digital revolution

Kesi Widjajanti
Universitas Semarang, Kota Semarang, Indonesia

ABSTRACT: The digital revolution is often discussed in terms of, and linked to, the company's ability to change in order to maintain its sustainable competitive advantage. The effects of the digital revolution forced changes in the company regarding its business operations. Its main problem is limited capacity of innovation. This article aims to explore how innovation can strategically improve efficiency and effectiveness business of a firm. A resource-based and market-based approach plays a role in the process of building innovation by taking internal and external factors into account. Innovation can be a source of sustainability in competitiveness.

Keywords: strategy, innovation, sustainability, digital

1 INTRODUCTION

Currently, Indonesia is trying to improve its competitiveness through various sectors considered reliable based on resource strength. This is because Indonesia has various resources that are well-known as having their own characteristics by expressing local cultures. However, in fact, the role of resources is not yet optimal as there are many obstacles, such as 1) lack of knowledge-related human resources, 2) lack of innovation capacity, and 3) lack of risk-taking attitudes in implementing new innovation. Such limitations have resulted in less innovation performance, leading to decreased competitiveness. Thus, to improve competitive capacity, companies need to make a change to more awareness on innovation and to understanding customer needs. The improvement of performance innovation will be achieved by developing innovation organizational and human capital (Widjajanti & Widodo, 2016).

Many studies related to performance of business firms focus on financial performance to create added value to maintain competitiveness. However, they mainly focus on profit and only a few studies focus on innovation performance and the sustainability of competitive advantage. Research on firms conducted by Widjajanti and Widodo (2016) discusses significant relationships in organizational innovation. This article aims to offer important contributions to understanding building innovation related to the digital revolution. It explains the improvement of capacity processes in creating innovation through market factors and resourced-based relationships. The success of business is not only determined by internal resources, but also by external factors such as marketing orientation with customers and suppliers. The existence of specific resources in a company is possibly related to improved innovation performance.

The merger of the resource-based view and the market-based view is expected to develop innovation capacity related to accelerated innovative behavior. This study focuses on specific resource factors indicated by existing knowledge of the relationship of capital human resources to innovative capacity indicated by the company's cooperation with suppliers and customers. The problem discussed is how business firms can create competitiveness through the role of innovation capacity to support innovative performance.

2 BUILDING INNOVATION: RESOURCE-BASED AND MARKET-BASED VIEW

The term "competitiveness" was initially introduced through the concept of comparative advantage by Ricardo in the eighteen century and has remained popular over the last three decades. A company's competitiveness is seen in the industrial, regional, or national level and is a combination of trade performance supporting internal and external capacity. Rostow (1999) says that competitiveness is ability to produce goods and services affecting international competition.

The aim of this article is to develop a strategic model of improving competitiveness mediated by the role of innovation. The combined resources-based view and market-based view approach is used to discuss indicators contributing to the ability of establishing cooperation with business partners. This perspective explains how innovation can create effectiveness in the market to increase quality of services and products. Initially, the concept of resource-based view and market-based view was developed by Makhija (2003) and Widjajanti (2015), emphasizing marketing collaboration factors that can provide complementary resources. The success of business can be seen from relationships showing commitment of marketing orientation. The approach of market-based view comes from mutualism in social network. Building innovation based on digital knowledge is a strategic approach that addresses the importance of performance innovation as the basis for competitive advantage. Previous studies find that competitive advantage created based on partnerships is an important resource to improve innovative performance.

Meanwhile, there is another perspective based on the resource-based view that says that the competitive advantage can be improved through the potential of internal resources. The resource-based view is an approach used for discussing resources and companies' capability to justify and predict external factors affecting competitive advantage. As mentioned by (Barney, 1991), Wenerfelt (2000), and Teece (1997), the resource-based view is an approach used by a company in creating its sustainable competitiveness based on resources. Previous researchers suggest that competitive advantage is created based on resources to improve performance. One of them is Ireland (2003), who says that resources that are valuable, rare, imperfectly imitable, and non-substitutable compared to the competitors', are important factors to improve competitiveness.

This article focus on strategic approaches that tend to consider innovation as the core of a competitive position. Some activities addressing competitive advantage require business strategies directed at a cooperative strategy compared to a competitive one. To survive in digital revolution, businesses should establish cooperation to expand their business networks. The company will be more powerful by establishing marketing collaboration with partnerships that can ease access to market and finance so that they can produce the products efficiently at a competitive price. Therefore, companies are not only required to provide better service but also to find new ways to be more efficient. Hence, there is a need to develop innovative capacity to create competitiveness.

3 ROLE OF RELATIONSHIP AND NETWORKING: INNOVATION STRATEGY TOWARD SUSTAINABLE COMPETITIVE ADVANTAGE

The strategy in improving innovative performance is an alternative solution for business sustainability. Company competitiveness in Indonesia can be created from specific resources of that stimulate innovation capacity. The definition of competitive advantage is various and depends on the approaches used. This article focuses on company competitiveness regarded as a result of a strategy combining the role of relationship as a support for maintaining market position benefiting the company for its sustainability. Partnership to support innovation performance is used for mediation to reach higher profits. In other words, cooperation is a strategy used for making development sustainable. Implementation from two perspectives of resource-based and market-based perspectives can create a strategy to respond to the digital revolution, reducing costs by optimizing companies' existing resources and expanding the market. The companies can get it from the relationship of input factors including innovation management and knowledge of human capital, which is difficult to gather.

This means that improving market share through resource capacity is not enough. Companies must establish cooperation with suppliers and customers to improve sales. A reputation for cooperation with business partners is also a resource for competitiveness affecting sales volumes. The integration of resource-based and factor market-based views is the basis for strategic approach to improve innovation performance. The company establishing partnerships is assumed to profit from the offered opportunities and integrate existing resources to improve competitiveness levels.

Increased competitiveness can be obtained by establishing wider relationships for business networking. Justifying the role of relationships in better market mobilization is the basis for creating strategies to optimize the competitive position. To survive in such fierce and tight competition, its necessary to maintain communication that makes it possible to create trust, which will in turn affect the network capability to facilitate beneficial market opportunities. The existing network can expand the field of cooperation, which can affect cost-sharing causing cost competitiveness. Further, the established relationship is a strategic way of improving the buyer-supplier relationship and is also related to stakeholders stimulating easy access for financial resources for company competitiveness. Having improved and easier financial access will automatically improve company capital to expand the business, leading to a better competitive position.

The owner and management of the company are expected to reorganize social capital to supporting the process of capacity improvement, which is dynamic and is related to information exchange from related partners to improve efficiency. It is argued that partnership is one of the best solutions to trigger competitiveness related to market situations. In fact, the existing market, adapted to conditions of resources and strategic actions, will be useful for gaining more profit. Likewise, internal factors such as existing knowledge and specific technical and design skills can make the relationship process successful. In mobilizing for business growth, organizational aspects such as management and group work with associations and clusters are vital to network development.

Ideally, all enterprises need to improve their competitiveness. To respond to the digital revolution, there is a need to establish relationships leading to effectiveness of production either in raw material provision or marketing. Improved company competitiveness in Indonesia is determined by creating mutual cooperation. The concept of a resource-based and market-based view can be the basis for strategic orientation in integrating and creating internal and relational resources.

The consideration in improving innovation capacity related to supplier and customers is called a "strategy driver" and it is an asset that can minimize external threats and cope with internal weaknesses. Innovation development can create access for complementary resources, reflecting strategic actions and opening more opportunities for marketing digital.

Therefore, relational factors are needed to formulate innovation strategies to improve business networking. it means that companies not only focus on social and knowledge capital, but also improve their capacity by developing business networking to market products and benefit from their partners. Company competitiveness based on internal and external orientation needs to be developed for business growth. Connections with partners also need to be nurtured to create profit so that business can be developed and can reach international markets.

4 CONCLUSION AND RECOMMENDATIONS

This article confirms that business firms that use existing knowledge of innovation as a specific resource can develop their networking and maintain a competitive advantage in the long run. Innovation capacity is an important internal resources to improve networking and the relationship between suppliers and customers.

The digital revolution can influence the increase of sales and profit, helping business firms operate more efficiently. Improved relationships play a role in the creation of specific resources for sustainable competitiveness. Companies act rationally to develop their competitiveness by not only considering their internal factors but also their marketing orientation as a process to develop their networks.

The best strategy to make the business sustainable is not directly determined by the condition of resources but it is also mediated by the ability to create wider relationships. Resource-based/market-based views offer a strategic alternative perspective for sustained company advantage. In the implementation of these perspectives, this study has managerial implications, meaning that the owners or the managers of the company need to develop capacity, especially the capacity of their employees in dealing with the customers and suppliers, such as knowing design trends, services, and products.

The recommendation to generate innovations that refine and reinforce the existing market, product, and service is not only determined by the quality of the resources but also by interactive process in the development of knowledge. The level of innovation needs to be increased, to answer digital revolution challenges that can have increased efficiency and quality. Marketing collaboration needs to be expanded to capture opportunities for a digital revolution to complement existing resources that can facilitate the success of innovation. Organizational innovation can be achieved by improving the quantity and quality of training and meeting with external institutions in order to open up to a better mindset and to support creativity. New ideas need to be injected into the business through the skills and knowledge of all existing personnel in the firm.

REFERENCES

Barney, J. (1991) "Firm resources and sustained competitive advantage." *Journal of Management*. Sage publications Sage CA: Thousand Oaks, CA, 17(1), pp. 99–120.

Gray, D. M. (2005) "The relationship of market orientation to business partnering performance." *ANZMAC 2005 Conference: Strategic Marketing and Market Orientation*, pp. 23–29.

Makhija, M. (2003) "Comparing the resource-based and market-based views of the firm: Empirical evidence from Czech privatization." *Strategic Management Journal*, 24(5), pp. 433–451.

Porter, M. E. (1998) "Competitive advantage, creating and sustaining superior performance, with a new introduction."New York, *The Free Press*.

Schroeder, R. G., Bates, K. A., and Juntila, M. A. (2002) "A resource based view of manufacturing strategy and relationship to manufacturing performance." *Strategic Management Journal*, 23, pp. 105–117.

Wernerfelt, B. (1984) "A resources based view of the firm." *Strategic Management Journal*, 5(2), pp. 171–180.

Wheelen, T. L., and Hunger, J. D. (2012) *Strategic Management and Business Policy Toward Global Sustaibanility*. Pearson Education, Inc: Prentice Hall, 13th ed.

Widjajanti, K. (2015) "Marketing collaboration and SME strategy implementation in Blora, Indonesia."*ASEAN Marketing Journal*, pp. 28–39.

Widjajanti, K., and Widodo. (2016) "Improving organizational innovation by human capital, knowledge sharing and organizational learning base." *International Business Management*, 10(9), pp. 1599–1609.

Zhou, I., Wu, W., and Luo, X. (2007) "Internationalization and the performance of born-global SMEs: the mediating role of social networks." *Journal of International Business Studies*, 38(4), pp. 673–690.

Facing Global Digital Revolution – Nirmala Arum Janie, Dwi Mulyaningsih & Wahyu Rachmawati (eds)
© 2020 Taylor & Francis Group, London, ISBN 978-0-367-33912-8

The use of internet media by micro banking Regional Owned Enterprises (ROEs) — evidence in Indonesia

Y. Kuntari
STIE Widya Manggala, Semarang, Indonesia

A. Chariri & T.J.W. Prabowo
Universitas Diponegoro, Semarang, Indonesia

Nurdhiana
STIE Widya Manggala, Semarang, Indonesia

ABSTRACT: Internet media plays an important role in communicating business activities, especially financial and non-financial performance. Organizations can utilize internet media to build transparency and accountability. The adoption of web-based technology is becoming a global trend. Websites are the best information technology today for exchanging information with the customers or clients easily, quickly, and without limitations of distance and time. The purpose of this study is to describe the use of internet media and its content by micro banking organizations owned by the Central Java, Indonesia, regional government. The population of this study were all micro banks owned by Central Java Regional Government, known as Perusda BPR BKK (Badan Kredit Kecamatan) and Perusda BPR BP (Bank Pasar). Technique of data acquisition was documentation and observation. The result of this research is identification of internet media usage along with micro banking content owned by local government.

1 INTRODUCTION

1.1 *Background*

Micro banking is understood as an instrument for self-empowerment by providing opportunities for the poor and common people to become agents of economic changes. Micro banking is considered as instrumental in reviving the people's economy because it is able to reach micro, small, and medium enterprises. Indeed, micro business has been viewed as less able to reach formal banking access. With the existence of micro banking then, access to capital can be possible for micro businesses, allowing for an increase in productivity of micro, small, and medium enterprises. Micro banking is established to cope with the obstacles of financing access to formal financial institutions. To overcome these obstacles, the community and government has established and developed many non-bank financial institutions that serve as business development and community empowerment, well known as microfinance institutions (Otoritas Jasa Keuangan, 2013).

In order to encourage the empowerment of micro business directly, in contact with the community, the regional government of Central Java established a micro banking business unit. Micro banking is the participation of regional government capital in Central Java, namely People's Credit Bank Regional Owned Enterprise (BPR BKK) and Market Bank Regional Owned Enterprise (Bank Pasar). These are intended to help micro entrepreneurs and the public who have limited access to formal banking. They provide services including loans or financing in micro-scale enterprises to members and communities, and savings management, as well as providing not-for-profit business development consulting services.

Internet media is a communication medium that can be used to provide information to stakeholders (investors, creditors, government, communities, and others), to allow them to make economic decisions related to their interests. This paper aims to describe internet media used by Central Java Regional Owned Enterprises in providing information and communicating with stakeholders, as well as the contents of published information.

2 LITERATURE REVIEW

2.1 *Micro banking*

Microfinance can be defined as the provision of both savings and credit financial services to micro enterprises (Waterfield & Duval, 1997). Another definition emphasizes the evolution of microfinance as an economic development tool to benefit low-income women and men, with savings, credit, insurance, and money-deposit services (Ledgerwood, 2000). In this sense, microfinance emphasizes development toward more modern banking services by increasing insurance services, as in the modern banking industry. It is found that clients of microfinance institutions are not just microenterprises seeking to finance their businesses (Christen, Lyman, & Rosenberg, 2003), they are micro clients who use microfinance to manage emergencies, acquire household assets, improve their homes, smooth consumption, and fund social obligations. Microfinance services are more concerned with customers' consumption habits (Christen, Lyman, & Rosenberg, 2003). In developing countries, such as Nigeria, microfinance institutions provide support for the economically active poor, low-income earners, and small- and medium-scale entrepreneurs. Micro banking is believed to be the driver of socioeconomic development (Nwabueze et al., 2013). With easy access to capital in micro banking, there will be more economic activity at the lower/grassroots level. In Uganda, the microfinance sector has both formal and informal elements. The formal form are either companies that are regulated under the banking laws or financial intermediaries that are not banks but regulated by the government (Ssewanyana, 2009). In Indonesia, microfinance institutions are regulated by the Financial Services Authority (OJK). Microfinance institutions are defined as financial institutions that are specifically established to provide their members and the public with business development services and community empowerment, either through loans or financing micro-scale enterprises, savings management, as well as not-for-profit business development consultancy services (Otoritas Jasa Keuangan, 2013). The goal of microfinance institutions is to improve access to micro-scale finances for the community, to provide assistance to empower the community, and to increase the community's income and welfare, especially for poor or low-income people.

In order to develop the economy of the community, local government in Central Java has established Regional Owned Enterprises. People's Credit Bank is a local company owned by the local government. Through the regional regulation (Perda) of Central Java Province, No. 3 of 2012 is a regulated Regional Owned Enterprise BPR BKK. They are located in the city/regency. Operationally, they can stand and open branches at the subdistrict level to get closer to serving the community.

2.2 *Internet media reporting*

It is undeniable that effective communication media play important roles in monitoring micro banking activities, encouraging growth, and realizing their potential. Communication media is certainly necessary so that stakeholders can access the required information.

Communication media experienced fairly rapid development associated with the emergence of internet technology, which is very helpful in the business world. Management can use internet technology as a communication medium to establish effective relationships with stakeholders. Information about the company, profile, products sold, and performance reports both financial and non-financial can be communicated. Thus, internet media, including social media, can be a means of communication between managers and investors/owners of capital.

The social media concept of the 21st century phenomenon was brought about by the advent of the Internet and the World Wide Web. Social media allows for direct communication and interaction between transmitters and receivers of information. It is argued that social media can be seen as using web-based technologies to transform and broadcast media monologues into social media dialogues (Kaplan & Haenlein, 2010). Communication and social media technologies are proven to provide information on microfinance schemes to the poor and low-income earners and micro, small- and medium-scale entrepreneurs, and information on how to access and use social media in airing their views on microfinance issues is essential (Nwabueze et al., 2013).

2.3 *Agency theory and signaling theory*

According to agency theory, agency problems can occur in a business organization. Management as agent and capital owner as principal have their own interests to improve their prosperity (Watts & Zimmerman, 1986). Management, of course, has more information than the principal. To bridge the asymmetry of such information, a reporting mechanism from management to owners is required.

In the context of signaling theory, management seeks to provide a good signal to the owners of capital (Musleh Al-Sartawi & Reyad, 2018), to ensure that management is maintained and supported by investors and other stakeholders. Information submitted by management to the owner of capital is important. Good management performance needs to be conveyed to the owners of capital either as a form of responsibility or as a medium of promotion.

3 METHODS

This research uses a quantitative descriptive approach, namely,documentary data gathered from internet media published by BPR BKK and BPR BP. Table 1 shows list of Regional Owned Micro Banks and the number of micro banks in the cities and districts. Based on the availability of data, we used 60 BPR offices at the district/city level, as well as in branches at the sub-district level. Documentation was accumulated by searching the internet media of the banks in Google, both their websites and other social media as described in Table 2.

Table 1. List of regional owned micro banks.

Regions	Number of micro banks	Regions	Number of micro banks
Magelang City	2	District of Pemalang	2
Pekalongan City	1	District of Demak, dst	1
Semarang City	2	District of Sragen	2
Surakarta City	-	District of Rembang	2
Tegal City	4	District of Wonogiri	1
Salatiga City	1	District of Sukoharjo	3
District of Banjarnegara	1	District of Klaten	4
District of Blora	2	District of Banyumas	-
District of Boyolali	3	District of Pekalongan	1
District of Sragen	2	District of Kudus	2
District of Batang	1	District of Jepara	2
District of Grobogan	1	District of Purbalingga	1
District of Kendal	1	District of Kr. Anyar	3
District of Wonosobo	2	District of Semarang	1
District of Brebes	1	District of Kebumen	2
District of Pati	2	District of Temangung	2
District of Purworejo	2	District of Purwokerto	1
District of Magelang	2		
Total	30		30

Table 2. The use of internet media by micro banks.

Internet Media (40)	Content of Information				
	Financial Reporting	Products/ Services	Bank Activities	Bank Profile	Others
A. Website (36)	21	32	27	29	9
B. Non Website (4)					
Social Media (Facebook and Tweeter) by both website and non-website	7	13	13	13	-
No internet (20)					

Source: Central Java Regional Owned Enterprises BPR BKK, 2017 (treated)

Table 2 describes the proportion of internet media usage and content usage. Internet media has been used by 40 BPR or 66.67% of micro banks: PD BPR BKK and BPR BP Central Java, and the remaining 20 BPR or 33.33% have not used internet media in communicating their business information. Its also indicates that 21 BPR (51.5%) utilized internet media to publish their financial reports, whereas the remaining 19 BPR (48.5%) use internet media for marketing (product information, services and activities). Interestingly, from the content of this information, BPR mostly used their website to publish information regarding bank product/services (32), bank profiles (29), and bank activities (27). The remaining information was financial reports and other publications.

4 DISCUSSION

The findings showed that not all Central Java Regional Owned Enterprises BPR BKK and BPR BP used internet media in communicating their activities. Indeed, some BPRs do not use internet media at all, even though, in the current era, the internet is the best media for a number of purposes, especially for business interests. In the business world, the internet is an inevitable necessity. Through social media, almost everyone is currently connected as social media shifted from entertainment to the world of work and business becoming an effective medium in business communication. The use of internet media is becoming popular because the ability to reach people and business at a geographical distance can be relatively difficult. About 13 micro banking use social media (Facebook and Twitter).

In regard to micro banks, the finding shows that micro banking utilized internet as communication media. This is because organizations are always looking for ways to have more value in the communities for the purpose of legitimacy. In fact, management trys to gain legitimacy in many ways by sending important signals concerning their activities. Publication of company profiles, activities, products, services, and financial performance can increase the value of the company in the eyes of stakeholders. The publication of financial statements in particular, can provide stakeholders with historical performance and insight into the company's future (Chariri, 2011). In terms of historical performance, annual reports can be used as a device to diagnose past events and to attribute praise or blame. Insight into the company's future can be used to see how capable management is of coping with future challenges and threats.

Annual reports, print media, and internet media are communication media that can be used by companies. Internet media becomes a necessity in a business organization to communicate between management to report both financial and non-financial performance to stakeholders. By using the Internet, organization builds transparency and accountability. The adoption of web-based technology is becoming a global trend in almost all organizations. The website is the best information technology today and are used by companies of various business purposes, including micro banking.

5 CONCLUSIONS

From the discussion, the following can be concluded about the use of internet media by BPR BKK and BPR BP:

1) Internet media users amounted to 40 out of 60 BPRs owned by Central Java Regional Owneds Enterprises (66.67%); the remaining 20 (33.33%) BPRs did not use internet media
2) Twitter, Facebook users: 14 BPRs; Website users: 36 BPRs
3) The use of website media: 21 BPRs use for the publication of financial statements, 32 BPRs for products and services, 27 BPRs for bank activities, 29 for company profiles, and 9 for others publications.

This shows that micro banks owned by Central Java government have not maximized their use of internet media to improve performance and need authorities, in this case OJK, have to provide guidance and coaching.

REFERENCES

Chariri, A. (2011) "Rhetorics in financial reporting: an interpretive case study," *Jurnal Akuntansi dan Keuangan*, 12(2), pp. 53–70.

Christen, R. P., Lyman, T. R., & Rosenberg, R. (2003) *Microfinance Consensus Guidelines: Guiding Principles on Regulation and Supervision of Microfinance*. CGAP and World Bank, Washington, DC.

Kaplan, A. M., & Haenlein, M. (2010) "Users of the world, unite! The challenges and opportunities of Social Media," *Business Horizons*. Elsevier, 53(1), pp. 59–68.

Ledgerwood, J. (2000) *Sustaining Banking with the Poor: Microfinance Handbook, An Institutional and Financial Perspective*, Washington, DC: The World Bank.

Musleh Al-Sartawi, A., & Reyad, S. (2018) "Signaling theory and the determinants of online financial disclosure," *Journal of Economic and Administrative Sciences*. Emerald Publishing Limited, 34(3), pp. 237–247.

Nwabueze, Chinenye, Nwabueze, Chizoba, & Egbra, O. (2013) "New communication technologies and microfinance banking in Nigeria: Critical role of the social media," *New Media and Mass Communication*, 15, pp. 12–17.

Otoritas Jasa Keuangan. (2013) *Lembaga Keuangan Mikro, Otoritas Jasa Keuangan.*

Ssewanyana, J. K. (2009) "ICT usage in microfinance institutions in Uganda," *The African Journal of Information Systems*, 1(3), p. 3.

Waterfield, C., & Duval, A. (1997) *CARE Savings and Credit Sourcebook*. CARE: Atlanta, GA.

Watts, R. L., & Zimmerman, J. L. (1986) *Positive Accounting Theory*. Prentice-Hall.

Facing Global Digital Revolution – Nirmala Arum Janie, Dwi Mulyaningsih & Wahyu Rachmawati (eds)
© 2020 Taylor & Francis Group, London, ISBN 978-0-367-33912-8

Aggressive financial reporting, boards of commissioners, and tax aggressiveness: An insight from Indonesia

A. Chariri, I. Januarti, E.N.A. Yuyetta & A.S. Adiwibowo
Universitas Diponegoro, Semarang, Indonesia

ABSTRACT: This study aims at investigating the effect of aggressive financial reporting and independent boards of commissioners on tax aggressiveness. using data from 200 annual reports of companies listed in the indonesia stock exchanges as the sample, which was then analyzed using a regression model, this study showed two main findings. as predicted, aggressive financial reporting and independent boards of commissioners significantly affect tax aggressiveness. companies involved in aggressive financial reporting tend to get involved in tax aggressiveness. interestingly, an independent board of commissioners also plays an important role in decreasing tax aggressiveness. our findings extend previous studies involving the possibility that financial reporting manipulation may be associated with taxation reporting manipulation.

1 INTRODUCTION

Tax aggressiveness has been considered a common phenomenon in countries that adopt self-assessment systems. However, no governments are able to rely solely on tax payers' consciousness of tax compliance (Slemrod, 2007), which may lead to tax aggressiveness. Frank, Lynch, and Rego (2009) claim that tax aggressiveness is concerned with manipulation of taxable income reported to tax authorities through tax planning activities, which may be legal, illegal or fall into a grey area (Chen et al., 2010). In fact, Shackelford and Shevlin, 2001, insist that companies may report financial income and taxable income differently, for example, because of different intentions in terms of financial and taxable income reporting.

Tax aggressiveness may correlate with aggressive financial reporting as companies report accounting information aggressively to stockholders by increasing income and, at the same time, report tax aggressively to fiscal authorities by decreasing taxable income (Frank et al., 2009). Thus, the relationship of aggressive financial reporting and tax aggressiveness have been seen as contentious issues in accounting literatures (see, for example, Frank et al., 2009; Hashim, Ariff, & Amrah, 2016; Lennox & Pittman, 2010; Nor, Ahmad, & Saleh, 2010). As claimed by agency theory, an agent tends to provide asymmetric information to the principles, especially investors (in the case of aggressive financial reporting) and government (in the case of aggressive tax reporting). Thus, an agent may provide asymmetric information to investors and government.

In the context of Indonesia, as part of tax reforms, the Indonesian government has adopted a self-assessment system since 1984. This system should make tax payers more honest and increase their willingness to comply with tax rules. However, the implementation of the system has led to opportunistic behavior and has been abused by tax payers (Tarjo & Kusumawati, 2006).

In regard to corporate governance, Indonesia adopts a two-tier systems: a board of directors (BoD) and a board of commissioners (BoC). The BoC is similar to an independent board of directors in a one-tier system of corporate governance. Referring to previous studies, corporate governance mechanisms, especially independent boards of directors, are effective in constraining earnings manipulation and tax aggressiveness (Chan, Mo, & Zhou, 2013; Kourdoumpalou, 2016; Richardson & Lanis, 2011).

Considering the above arguments, this study aims to find empirical evidence on the relationship between aggressive financial reporting, independent BoCs, and tax aggressiveness. This research is expected to extend the scope of previous studies on aggressive financial reporting, corporate governance, and tax aggressiveness by considering the business environment in emerging markets that implement self-assessment systems as have been implemented since 1984. These results should be important to regulators such as the Indonesia Financial Service Authority and Tax Office, and other stakeholders who have a vested interest in understanding the interaction of financial and tax reporting decisions

2 HYPOTHESIS FORMULATION

2.1 *Aggressive financial reporting and aggressive tax reporting*

Borrowing agency theory claims on the self-interest proposition, agents may create "downward manipulation of taxable income through tax planning activities to reduce tax that should be contributed to the government" (Frank, Lynch, & Rego, 2009). Conversely, agents may create "upward earnings management that may or may not be within the confines of GAAP" for the purpose of getting bonuses or to show the owners that their performance is outstanding (Frank et al., 2009).

In addition, firms committing to earning management activities usually present lower discretionary current accruals in the year before the tax rate is reduced, and present higher discretionary current accruals after the tax rate is decreased (Guenther, 1994; Roubi & Richardson, 1998). If so, we expect a relationship between aggressive financial reporting and tax aggressiveness as follows.

H1: The more aggressive companies are in reporting higher accounting income, the more aggressive the companies are in reporting lower tax

2.2 *The independent board of commissioners and tax aggressiveness*

Agency theory claims that corporate governance plays an important role in deterring opportunistic behavior of an agent in financial reporting and tax reporting. Borrowing arguments of previous studies, accounting scholars believe that financial-statement fraud, for example, is influenced by the composition of the board members (Beasley, 1996), particularly independent members.

Previous studies conclude that independent boards of directors are effective in constraining earnings manipulation and accounting frauds (Beasley, Carcello, & Hermanson, 1999; Beasley et al., 2000; Carcello & Nagy, 2004; Fanning & Cogger, 1998; Fich & Shivdasani, 2007; Klein, 2002; Rezaee, 2005; Titus, Heinzelmann, & Boyle, 1995). These findings could be also relevant to tax aggressiveness issues as agents prefer to report lower taxable income to tax authorities. Therefore, we propose the following hypothesis:

H2. Independent boards of commissioners negatively influence tax aggressiveness

3 RESEARCH METHOD

Population of this study consisted of all companies listed on the Indonesia Stock Exchanges in the years 2013–2016. Data were collected from annual reports published on www.idx.co.id. Based on the availability of data, we got 53 companies but unfortunately three companies possessed negative pre-tax profit. Thus, we employed 200 data points (50 companies with four years' annual reports).

Aggressive financial reporting is measured by using the modified Jones model as described by Dechow, Sloan, and Sweeney (1995). Tax aggressiveness is based on effective tax rate: the ratio of tax expense to pre-tax income (Chan et al., 2013; Chen et al., 2010; Hashim et al., 2016; Putri, Rohman, & Chariri, 2016). Lower effective tax rates indicate higher tax aggressiveness (Ariff & Hashim, 2014; Chen et al., 2010; Hashim et al., 2016). The independent board of commissioners (IB) is measured by the percentage of independent members to total members of the BoC as stated on the annual reports (Beasley, Carcello, & Hermanson, 1999; Darmadi & Sodikin, 2013; Post, Rahman, & McQuillen, 2015). Data are then analyzed using multiple linear regression based on the following model:

$$TA = \alpha + \beta_1 AFR + \beta_2 IB + \varepsilon \quad (1)$$

Where α is intercept; ß1 shows regression coefficient; AFR represents Aggressive Financial Reporting, TA shows Tax Aggressiveness; IB reflects Independent Board of commissioners; and ε is errors.

4 FINDINGS AND DISCUSSION

This study aims to find empirical evidence of the relationship between aggressive financial reporting, independent boards of commissioners, and tax aggressiveness. The descriptive statistics of empirical data, as seen in Table 1, indicated that tax aggressiveness (TA) is relatively moderate with a mean score of 0.279. Meanwhile aggressive financial reporting (AFR) can be seen from discretionary accrual earnings management with mean score of 0.000. This means that the companies used in the sample are relatively less aggressive in managing their financial reporting even though they tend to increase their earnings. From the perspective of corporate governance (board of commissioner), it can be seen that the mean score of independent members (IB) of the BoC was 35.8% of total members of the boards. This implies that the number of independent members of the BoC is still low.

The results of regression model, as indicated in Table 2, shows that all hypotheses are supported by empirical data. Indeed, aggressive financial reporting significantly influenced tax aggressiveness ($p = 0.014$, less than 0.50). Moreover, independent board of commissioners significantly affected tax aggressiveness ($p = 0.026$, less than 0.05).

The findings indicated some interesting insights. First, aggressive financial reporting, which is measured by discretionary earnings management, significantly affects tax aggressiveness. The descriptive statistics shows that the level of earnings management, as a proxy of aggressive financial reporting, of the sample companies is relatively low and tax aggressiveness is moderate. This means that Indonesian companies committed to aggressive financial reporting

Table 1. Descriptive statistics.

Variable	N	Minimum	Maximum	Mean	Std. Deviation
TA	200	–0.088	1.456	0.279	0.144
AFR	200	–0.517	0.382	0.000	0.122
IB	200	0.142	0.800	0.358	0.123

Table 2. Results of regression (dependent= tax aggressiveness).

	Stand. Coeff	t	p-value	Note
(Constant)	0.344	11.180	0.000	-
AFR	–0.203	–2.480	0.014*	H1 Supported
IB	–0.182	–2.240	0.026*	H2 Supported

N = 200; F(2, 197)=5.25; Prob>F = 0.0060; Adj-R2=0.041
* Significant at 5%

and tax aggressiveness favor employment of the common planning technique rather than using more aggressive planning techniques as mentioned by Frank et al., 2009. Furthermore, the negative relationship of aggressive financial reporting and tax aggressiveness supports previous studies that companies prefer reporting higher accounting income to shareholders and lower taxable income to tax authorities (Roubi & Richardson, 1998; Shackelford & Shevlin, 2001).

Second, from the corporate governance perspective, the finding shows that independent boards of commissioners significantly affected tax aggressiveness. This finding supported previous claims that the boards of commissioners, especially independent members of BoCs, play significant roles in mitigating the agency problems of publicly listed companies due to the separation of ownership and management. The finding supports previous studies on the effective role of independent boards of directors (independent members of the BoC) in constraining unethical behavior of agents, such as their involvement in earnings manipulation and accounting frauds (Beasley et al., 1999, 2000; Carcello & Nagy, 2004; Fanning & Cogger, 1998; Fich & Shivdasani, 2007; Klein, 2002; Rezaee, 2005; Titus et al., 1995)

5 CONCLUSION

The empirical findings indicated some interesting insights. First of all, the level of aggressive financial reporting of Indonesian companies is relatively low, whereas tax aggressiveness is moderate. It is also found that the proportion of independent members in the BoC structure is low. The findings infer that the companies prefer to implement the common planning technique rather than using more aggressive planning techniques when they deal with such aggressiveness.

Second, the empirical data supported the first hypothesis that aggressive financial reporting significantly influenced tax aggressiveness. This means that companies listed in the Indonesia Stock Exchange prefer to report higher income to stockholders and at the same time favor reporting lower taxable income to tax authorities. Self-assessment systems provide companies with the opportunity to use tax rules to report tax aggressively.

Third, the empirical finding also supports the hypothesis that independent BoC negatively influence tax aggressiveness. The more independent members on the BoC, the less aggressive the companies are in managing their taxable income. Indeed, independent members of the BoC prevent companies reporting less taxable income than what should be reported.

Findings of this study provide us with some contributions. First, our findings extend previous studies involving the possibility that financial-reporting manipulation may be associated with taxation-reporting manipulation. Thus, we should analyze financial reporting comprehensively by including the taxation perspectives, and vice versa. This is because decisions regarding financial reporting and taxation are related, in that tax may affect income and the cash position of a company. Moreover, this study extends the scope of previous studies on aggressive financial reporting and tax aggressiveness to consider the business environments in emerging markets. Indeed, accounting scholars should consider contextual factors (such as politics, culture, and laws) when researching aggressive accounting and tax aggressiveness. Finally, accounting regulators need to consider the effect of accounting choices on the opportunistic behavior of agents for the purpose of reducing aggressive financial reporting and tax aggressiveness.

REFERENCES

Ariff, A. M., and Hashim, H. A. (2014) "Governance and the value relevance of tax avoidance." *Malaysian Accounting Review*, 13(2), pp. 87–108.

Beasley, M. S. (1996) "An empirical analysis of the relation between the board of director composition and financial statement fraud." *The Accounting Review*, 71(4), pp. 443–465.

Beasley, M. S., Carcello, J. V., and Hermanson, D. R. (1999) "Fraudulent financial reporting: 1987–1997: An analysis of U.S. public companies research." *The Committee of Sponsoring Organizations of the Treadway Commission (COSO)*.
Beasley, M. S., Carcello, J. V., Hermanson, D. R., and Lapides, P. D. (2000) "Fraudulent financial reporting: Consideration of industry traits and corporate governance mechanisms." *Accounting Horizons*, 14(4), pp. 441–454.
Carcello, J. V., and Nagy, A. L. (2004) "Client size, auditor specialization and fraudulent financial reporting." *Managerial Auditing Journal*, 19(5), pp. 651–668.
Chan, K. H., Mo, P. L. L., and Zhou, A. Y. (2013) "Government ownership, corporate governance and tax aggressiveness: Evidence from China." *Accounting & Finance*, 53(4), pp. 1029–1051.
Chen, S., Chen, X., Cheng, Q., and Shevlin, T. (2010) "Are family firms more tax aggressive than non-family firms?." *Journal of Financial Economics*, 95(1), pp. 41–61.
Darmadi, S., and Sodikin, A. (2013) "Information disclosure by family-controlled firms: The role of board independence and institutional ownership." *Asian Review of Accounting*, 21(3), pp. 223–240.
Dechow, P. M., Sloan, R. G., and Sweeney, A. P. (1995) "Detecting earnings management." *The Accounting Review*, 70(2), pp. 193–225.
Fanning, K. M., and Cogger, K. O. (1998) "Neural network detection of management fraud using published financial data." *Intelligent Systems in Accounting, Finance and Management*, 7(1), pp. 21–41.
Fich, E. M., and Shivdasani, A. (2007) "Financial fraud, director reputation, and shareholder wealth." *Journal of Financial Economics*, 86(2), pp. 306–336.
Frank, M. M., Lynch, L. J., and Rego, S. O. (2009) "Tax reporting aggressiveness to aggressive and its relation financial reporting." *The Accounting Review*, 84(2), pp. 467–496.
Guenther, D. A. (1994) "Earnings management in response to corporate tax rate changes: Evidence from the 1986 Tax Reform Act." *The Accounting Review*, 69(1), pp. 230–243.
Hashim, H. A., Ariff, A. M., and Amrah, M. R. (2016) "Accounting irregularities and tax aggressiveness." *International Journal of Economics, Management and Accounting*, 1(1), pp. 1–14.
Klein, A. (2002) "Audit committee, board of director characteristics, and earnings management." *Journal of Accounting and Economics*, 33(3), pp. 375–400.
Kourdoumpalou, S. (2016) "Do corporate governance best practices restrain tax evasion? Evidence from Greece." *Journal of Accounting and Taxation*, 8(1), pp. 1–10.
Lennox, C., and Pittman, J. (2010) "Big five audits and accounting fraud." *Contemporary Accounting Research*, 27(1), pp. 209–247.
Nor, J. M., Ahmad, N., and Saleh, N. M. (2010) "Fraudulent financial reporting and company characteristics: Tax audit evidence." *Journal of Financial Reporting and Accounting*, 8(2), pp. 128–142.
Post, C., Rahman, N., and McQuillen, C. (2015) "From board composition to corporate environmental performance through sustainability-themed alliances." *Journal of Business Ethics*, 130(2), pp. 423–435.
Putri, A., Rohman, A., and Chariri, A. (2016) "Tax avoidance, earnings management, and corporate governance mechanism (an evidence from Indonesia)." *International Journal of Economic Research*, 13(4).
Rezaee, Z. (2005) "Causes, consequences, and deterence of financial statement fraud." *Critical Perspectives on Accounting*, 16(3), pp. 277–298.
Richardson, G., and Lanis, R. (2011) "Corporate social responsibility and tax aggressiveness." *Journal of Accounting and Public Policy*, 31(1), pp. 86–108.
Roubi, R. R., and Richardson, A. W. (1998) "Managing discretionary accruals in response to reductions in corporate tax rates in Canada, Malaysia and Singapore." *The International Journal of Accounting*, 33(4), pp. 455–467.
Shackelford, D. A., and Shevlin, T. (2001) "Empirical tax research in accounting." *Journal of Accounting and Economics*, 31(1), pp. 321–387.
Slemrod, J. (2007) "Cheating ourselves: The economics of tax evasion." *The Journal of Economic Perspectives*, 21(1), pp. 25–48.
Tarjo, T., and Kusumawati, I. (2006) "Analisis Perilaku WP Orang Pribadi terhadap Pelaksanan Self Assessment System: Suatu Studi di Bangkalan (Analysis of tax payers' behavior on the implementation of self assessment systems)." *Jurnal Akuntansi dan Auditing Indonesia*, 10(1), pp. 101–120.
Titus, R. M., Heinzelmann, F., and Boyle, J. M. (1995) "Victimization of persons by fraud." *Crime & Delinquency*, 41(1), pp. 54–72.

Facing Global Digital Revolution – Nirmala Arum Janie,
Dwi Mulyaningsih & Wahyu Rachmawati (eds)
© 2020 Taylor & Francis Group, London, ISBN 978-0-367-33912-8

The importance of alumni feedback in the curriculum evaluation to improve the competencies of students of the undergraduate accounting program

Tri Jatmiko Wahyu Prabowo & Maal Naylah
Universitas Diponegoro, Semarang, Indonesia

ABSTRACT: This study aims to obtain views and feedback from important stakeholders, in particular, alumni, regarding the curriculum of the Diponegoro University undergraduate accounting program with an eye to improving graduate competencies. Grounded theory is used to organize, verify, interpret, and conceptualize data. This research is categorized as qualitative research that underlies the interpretive paradigm. Alumni provide valuable insights because they are able to illustrate the expected competencies from a "real world" perspective. This study highlights at least two valuable insights from the alumni: (1) IFRS is compulsory at the introductory level of the course, and then advanced IFRS must be the responsibility of accounting firms and companies to teach; (2) This study convincingly shows that undergraduate accounting programs should be able to provide high quality human resources (employees) who will not only work in accounting firms as an auditor.

Keywords: alumni, accounting graduate competency, accounting curriculum, stakeholders, undergraduate program

1 INTRODUCTION

For the past forty years, many criticisms have been expressed by professionals, academics, and graduates that accounting programs fail to deliver graduates with the necessary competencies required by the accounting profession and other users in the modern technology business environment (Bui & Porter, 2010). This failure demanded reforms in accounting education (Brown & McCartney, 1995; Thompson, 1995). There is an indication that the failure comes from the educational curriculum. Armitage (1991) identifies that there are differences in perspectives between accounting practitioners and accounting academicians regarding competencies that accounting graduates should possess, and several suitable programs for developing those competencies. Furthermore, Carr, Chua, and Perera (2006) state that issues in accounting education are caused by (i) inadequate attention given to curriculum development; (ii) limited involvement of stakeholders in solving education issues.

The undergraduate accounting program can be influenced not only by the requirements of professional associations, but also by interested parties, such as businesses, professionals, alumni, managerial universities, and communities (Stout et al., 2004). With the escalating demand for public accountability in the past few decades, there has been an increase in support for involving the stakeholders in the design of the accounting curriculum to improve graduate competencies. However, the literature of so-called stakeholder views generally has only a narrow view, focusing primarily on student perceptions (Carr, Chua, & Perera, 2006).

Some scholars identify alumni as important stakeholder groups. Stout et al. (2004) and Stout and West (2004) provide valuable input about the strengths and weaknesses of educational experiences and processes. Their feedback will be beneficial in designing an accounting program curriculum. Alumni are in a unique position to provide valuable feedback in designing curriculum for competency enhancement.

There is considerable demand from several parties interested in the university, such as professional associations, practitioners and even entrepreneurs in the corporate sector, to reform accounting education to produce competent accounting graduates who have added value for professional work (Carr et al., 2006). In response to the demand, the Undergraduate Accounting Program of Diponegoro University established a new curriculum, which considered two approaches, namely the Indonesian National Qualifications Framework and International Education Standards (IES) in 2016. In relation to the professional competencies of accounting graduates, IES established study programs that are compulsory for graduates to enter the professional level: (i) Financial Accounting and Reporting, (ii) Management Accounting, (iii) Finance and Financial Management, (iv) Taxation, (v) Audit and Guarantees, (vi) Governance, Risk Management And Internal Control, (v) Business Law and Regulation, (vi) Information Technology, (vii) Business and Organizational Environment, (viii) Economics, (ix) Business Strategy and Management.

This paper supports the recognition that there are important characteristics of the accounting curriculum that should be taken into account in designing an undergraduate accounting program. The important characteristics may emerge from the alumni. This paper intends to present the results of an empirical study that aims to ascertain the perspective of alumni as a stakeholder group, in particular, their perspective on the issue of improving graduate competence to meet stakeholder expectations.

2 RESEARCH METHOD

This study is qualitative research using grounded theory in organizing, verifying, interpreting, and conceptualizing data. Originally, grounded theory was a research method used to construct theoretical propositions. However, we only borrow grounded theories in analyzing data, not for developing theories. According to Strauss and Corbin (1998), grounded theory is produced by continuous interaction between data collection, analysis, and report writing based on data obtained from interviews and field studies or other sources. In analyzing data, researchers identify themes or dimensions that emerge from data collected from interviews, and use these to identify the key themes and develop relationships between themes.

We interviewed 40 alumni, consisting of 10 public accountants, 10 managers or internal auditors in the company, 10 government auditors, and 10 government officers. Before the interviews, we selected interviewee candidates from the list of alumni that was maintained by the Section of Alumni Tracing, Faculty of Economic and Business, Diponegoro University. From the list, we chose 60 candidates who have worked more than for years. We contacted the candidates by phone to ascertain their willingness and availability for interviews. Forty alumni agreed to be interviewed.

Interviews were conducted at the interviewee's office and lasted more than two hours. In the beginning, the interviewees were presented the current curriculum in the undergraduate accounting program of Diponegoro University and were asked what they thought about the course lists. We had already developed a list of questions, however, we often let the interview flow naturally without following the list. All interviews were conducted in the Indonesian language, recorded, and then transcribed.

From transcribed interviews, we continued to integrate into the preliminary model based on the themes that we obtained from interviews. Each theme was labeled with a code, which was open code. We identified more than 200 themes, which were labeled open code, for instance like innovative and creative, confident, accounting techniques, English communication. The process of developing open codes was conducted in conjunction with the interview process. The next step in data analysis was the development of axial codes. Weaknesses, increased competencies, core courses, and peripheral courses are examples of themes that we gave axial codes. There was no clear boundary between the two processes, open and axial codes. Sometimes, relationships among themes (axial coding) emerged accidentally when looking for descriptions of certain themes (open coding) and vice versa. After the axial coding process, the next step in data analysis in grounded theory was the refinement of themes, which produced selective code that provided an overview of the relationships among themes. We identified more than 30 themes, which consisted of accounting and non-accounting firms, professional work, IFRS courses, and communication skills. In the final process, we validated the findings through the final interview.

3 FINDINGS AND DISCUSSIONS

The skills to apply accounting and communication techniques (oral and written) are two competencies that almost all interviewees (34 persons) expect accounting graduates to possess. This finding is consistent with research findings (e.g., Carr, Chua, & Perera, 2006), which state that good communication skills are the most important competencies for accounting graduates. Interviewees also highlighted problem solving as an important skill.

Most interviewees (37 persons) emphasized communication in English. Lack of English proficiency and self-confidence are the major flaws that must be tackled by curriculum design. Instead of improving technical courses, they recommended that all resources should encourage improvement of soft skills in English proficiency and character development in increasing graduates' self-confidence.

Accounting and financial reports have been emphasized by interviewees (28 persons) as the most important courses in the technical aspects of accounting in this study. Management accounting also received considerable support from the interviewees (25 persons) especially from interviewees who did not work in accounting firms. Some interviewees recommended that graduates should pass the advanced level, and complete at least two of the three specification courses in the field: accounting and financial reports, management accounting, and audits and assurance. However, some interviewees (27 persons) assessed the audit courses as the least important course in this study, which might reflect arguments often rasied in the literature (Allen, 1999), namely, that audits should be the part of professional training programs rather than undergraduate courses.

Furthermore, the interviewees (31 persons) recommended that the direction of information technology should be improved to anticipate the latest major changes in technology: the 4.0 digital revolutions. Interviewees suggested that the lecturers of the undergraduate program should no longer operate in the seller's market with accounting firms who are the dominant consumers of their products. The market has changed. Findings assure us that undergraduate accounting programs should deliver high-quality human resources, not merely workers in accounting firms as auditors.

The field of taxation, which is more technical, is still important for alumni working in accounting firms. There is evidence that economics, finance and financial management, governance, risk management, and internal control are equally important.

In contrast to the adoption of the implementation of accounting techniques, most interviewees (33 persons), especially those who do not work in accounting firms, did not mention International Financial Reporting Standards (IFRS) as important knowledge and skills that must be acquired by accounting graduates. The recent adoption of IFRS is still considered by some interviewees as brand new and difficult to implement. They considered it enough that IFRS is taught at the introductory level. Then, the advanced levels of IFRS must be the responsibility of accounting firms and companies to teach.

Minor specializations for undergraduate accounting programs were not prioritized by the interviewees (24 persons). They prefer that accounting graduates have the flexibility to pursue jobs that are in line with their passion. Most interviewees speculated that minor specialties would limit the choices of accounting graduates in hunting jobs.

Interviewees (30 persons) criticized the curriculum of the undergraduate accounting program as consisting of too many unnecessary programs that do not support graduate competencies, such as Ideology of Pancasila, Basic Culture of Indonesia, and Religion. The unnecessary courses make students take too many courses each semester. As a result, the students were overloaded throughout the semester. Also, the method used to teach accounting is very technical and mechanical and does not stimulate students to be creative and innovative. Then, this has become a significant issue as soft skills must be taught to accounting graduates.

4 CONCLUSIONS

We found that each alumni group requires different competencies. It is very important to listen to all the "voices," including the voices of the alumni. Then, if it is applicable, we try to accommodate the input in curriculum development. The alumni group provided valuable feedback not only because they have completed the program but also because they have the privilege of experiencing "professional work." It is already mentioned that involving the stakeholders in designing the curriculum is very important. Lecturers should adopt a value-added approach in the curriculum-designing process. The stakeholders' perspectives are very useful in enhancing value-added competencies.

In conclusion, in the process of curriculum development, the issues of increasing competencies are far more complicated than can be solved by just relying on a particular market or group in the market. However, listening to feedback and the recommendations of the alumni is very valuable in understanding what problems have been faced by graduates of an accounting education.

REFERENCES

Allen, W. (1999) "The future of accounting education." *Pacific Accounting Review*. MCB UP Ltd, 11(1/2), pp. 1–7.

Armitage, J. L. (1991) "Academicians' and practitioners' views on the content and importance of the advanced financial accounting course." *Journal of Accounting Education*. Elsevier, 9(2), pp. 327–339.

Brown, R. B., and McCartney, S. (1995) "Competence is not enough: Meta-competence and accounting education." *Accounting Education*. Taylor & Francis, 4(1), pp. 43–53.

Bui, B., and Porter, B. (2010) "The expectation–performance gap in accounting education: An exploratory study." *Accounting Education: An International Journal*. Taylor & Francis, 19(1–2), pp. 23–50.

Carr, S., Chua, F., and Perera, H. (2006) "University accounting curricula: The perceptions of an alumni group." *Accounting Education: An International Journal*. Taylor & Francis, 15(4), pp. 359–376.

Stout, D. E., et al. (2004) "A descriptive account of the development and implementation of an innovative graduate accounting program." in *Advances in Accounting Education* Teaching *and* Curriculum Innovations. Emerald Group Publishing Limited, pp. 249–272.

Stout, D. E., and West, R. N. (2004) "Using a stakeholder-based process to develop and implement an innovative graduate-level course in management accounting." *Journal of Accounting Education*. Elsevier, 22(2), pp. 95–118.

Strauss, A., and Corbin, J. (1998) *Basics of Qualitative Research Techniques*. Sage Publications: Thousand Oaks, CA.

Thompson, P. J. (1995) "Competence-based learning and qualifications in the UK." *Accounting Education*. Taylor & Francis, 4(1), pp. 5–15.

*Facing Global Digital Revolution – Nirmala Arum Janie,
Dwi Mulyaningsih & Wahyu Rachmawati (eds)*

Entrepreneurial intention amongst undergraduates: Tendency and boosters to start-ups

I.A. Majid, A.R. Yunus & S.W.M. Yusof
FPTT, Universiti Teknikal Malaysia, Melaka, Malaysia

ABSTRACT: Awareness of entrepreneurship in terms of intention to set up one's own business and self-employment, in addition to common recognition of the usefulness of enterprise skills when competing in the business world, were among the generally accepted reasons behind the initiative of promoting enterprise education. Specifically, enterprise education is the process of training individuals so that they can recognize business opportunities, and have the knowledge and related skills to respond through setting up a business venture. This article explores factors that encourage undergraduate students to decide to become entrepreneurs upon finishing their studies at the university. The results revealed that "attracted by success of other entrepreneurs," seems to be the most agreed-upon factor that influences students to become entrepreneurs upon graduation.

1 INTRODUCTION

The Global Entrepreneurship Monitor's (GEM Consortium, 2017) confirms the impacts made by the complex and diverse entrepreneurial activities around the globe based on start-up initiatives, economic value, and contribution to countries' sustainability and economic development. In general, entrepreneurship has been perceived as a major promoter of continuous growth and development in the economy of many countries as it not only creates employment, but also intensifies various economic activities such as market growth, knowledge transfers, employment, and innovation (Nowiński & Haddoud, 2019; Aman et al., 2011).

Awareness of entrepreneurship in terms of intention to set up one's own business and self-employment, in addition to common recognition of the usefulness of enterprise skills when competing in the business world, were among the generally accepted reasons behind the initiative of promoting enterprise education. Specifically, enterprise education is the process of training individuals so that they can recognize business opportunities, and have the knowledge and related skills to respond through setting up a business venture. However, there is a constant argument about whether or not we can actually teach and train students to become entrepreneurs (Fiet, 2001). This dispute is also proposed to begin from the debate on whether entrepreneurs are born or made.

Pulka, Rikwentishe, and Ibrahim (2014) posited that personality and environmental forces were the major factors that mold individuals' attitude and decision whether they will or will not become involved in entrepreneurial activities. Pekkala Kerr, Kerr, and Xu (2017) conceded the main factors that may influence the inclination for entrepreneurship are the propensity for taking calculated risks (Brockhaus, 1980), internal locus of control (Brockhaus, 1985), the need for achievement (McClelland, 1961), and the desire for personal control (Greenberger & Sexton, 1988). Potential factors that may determine entrepreneurial behavior could also include the individual's working experience or career history (Storey, 1982; Ronstadt, 1988). Other factors include a person's gender (Buttner & Rosen, 1989; Kolvereid et al., 1993), level of education (Storey, 1982), family background (Scott & Twomey, 1988; Matthew & Moser, 1995), and race (Aldrich & Waldinger, 1990). Thus, this study also is intended to investigate the profile of a potential entrepreneur.

2 ENTERPRISE AS A CAREER

Researchers in entrepreneurship commonly study factors that motivate individuals to venture into business start-ups rather than focusing on career advancement and progression. Success in business ventures relies mainly on a graduate's business skills (GEM Consortium, 2017). Entrepreneurship education and training are designed to stimulate students to go for self-employment upon graduation. Through this course, students are exposed to various ways and options to begin business ventures (Fatoki, 2010; Katundu & Gabagambi, 2016).

Trait theory attempts to determine common relationships among entrepreneurs that tie them together as a group. These include self-belief, a high need for achievement, independence, and a propensity to take risks (McClelland, 1961, and Rotter, 1966). However, no single trait has been proven to be exclusively entrepreneurial and trait measurement is subject to continuous debates.

The social development approach acknowledges that decision-makers frequently have access to limited data and are susceptible to external limitations and influences at different levels. Factors such as risk, family influences, prior education and training, and perceived job opportunities may be the common variables related to this situation (Gibb & Ritchie, 1982). However, the model does not account for young entrepreneurs nor, importantly, other influences such as peer groups and location.

The structure opportunity model rejects the two previous approaches and claims that they provide insufficient emphasis on social factors such as family background, neighborhood, school, peer group, and general work situation. The latter imperative demand-side factor infers that career choice is affected by employers' needs and the by and large financial and work climate. Curran (1996) also recommends the prevailing family attitudes, friends, and neighbors perpetuate a young person's social position so that esteem structures and states of mind influence their awareness.

Factors that may lead to individuals' intention to become an entrepreneur include individual or psychological components, and social and economic situations. These socialization backgrounds and experiences may become determining factors for an entrepreneurial career. For example, Dalton and Holloway (1989) found that many successful entrepreneurs had received significant responsibilities at a young age, even to the extent of some starting entrepreneurial ventures. Effective training and education on running a business can also push some people toward such a career.

3 METHODOLOGY

The population under study consists of students from two of Malaysia's PHLIs: Universiti Teknikal Malaysia (UTeM) and Universiti Teknologi Malaysia (UTM). A total of 1,000 samples were selected from these two universities through simple random sampling method. These two universities were chosen because members of the research team for this study are lecturers in these universities. Therefore, convenience for data collection process and administration of the study was why these two universities were chosen as sample for the study.

Primary data was collected through self-administered questionnaires, which were distributed to the respondents by the members of the research team from both universities. This method is perceived as most appropriate as it allows for a large sample spread over a wide area to be surveyed and enables quick response. It also avoids personal bias.

4 RESULTS

4.1 *Tendency for business start-up amongst university students*

Table 1 presents the frequency counts and percentages of respondents' tendency for business start-up amongst university students. The result shows that 161 students (25%) are very enthusiastic to start their own business and 301 students (46%) are enthusiastic to start businesses. On the other hand, 171 students (26%) are not very enthusiastic to start up businesses and 23 students (4%) are not enthusiastic at all to start up business.

4.2 *Boosters to start-up as perceived by university students*

This section provides descriptive analyses of the respondents' responses based on the various items assumed and perceived as boosters to business start-up as shown in Table 2.

From the list of suggested factors that might be perceived as boosters to business start-up among the students, first on the list based on the mean score of the items is "attracted by the success of other entrepreneurs," while last on the list is "encouraged by friends and relatives."

The analysis found that 324 respondents (49.4%) agreed and a further 250 respondents (38.1%) strongly agreed that they are attracted by the success of other entrepreneurs. On the other hand, only 14 (2.1%) of the respondents stated that they strongly disagree and another 68 (10.4%) of the respondents disagree with this statement. From this analysis, it can be implied that the success of other entrepreneurs has attracted students to become entrepreneurs. Arranged in a descending order the factors related here are as follows: "attracted by the success of other entrepreneurs"; "believed that entrepreneurship will provide one with the opportunities to accumulate wealth and profit"; "wanted to be self-employed"; "believed that the opportunities of employment are limited", "wanted to use own expertise, "could sense the business opportunities"; "encouraged by government's policies and incentives"; and encouragement by relatives and friends." Ironically, despite the wide range of financial assistance provided by the Malaysian government for the establishment and development of SMEs in Malaysia as forwarded earlier in this article, the related item "encouraged by government's policies and incentives," is among the lowest factors perceived by students as boosting their business start-up tendency. This scenario perhaps can be considered as reflecting the ignorance among the students of the abundance of opportunities and assistance provided by the Malaysian government.

Table 1. Frequency count and percentages of respondents' level of enthusiasm to start up businesses.

Enthusiasm to start up businesses	N sample	Percentage
Very enthusiastic	161	24.5
Enthusiastic	301	45.9
Not very enthusiastic	171	26.1
Not enthusiastic at all	23	3.5
Total	656	100.0

Table 2. Descriptive statistics based on responses to items regarded as boosters to becoming an entrepreneur.

Variable	Variable value	No. of cases	% of cases	Mean	Standard Deviation
I believed the opportunities of employment are limited	Strongly disagree	16	2.4	3.14	0.71
	Disagree	76	11.6		
	Agree	363	55.3		
	Strongly agree	201	30.6		
I want to be self-employed	Strongly disagree	12	1.8	3.14	0.74
	Disagree	102	15.5		
	Agree	321	48.9		
	Strongly agree	221	33.7		
I want to use my own expertise	Strongly disagree	21	3.2	2.99	0.72
	Disagree	111	16.9		
	Agree	375	57.2		
	Strongly agree	149	22.7		
I believed entrepreneurship will provide me with the opportunities to accumulate wealth and profit	Strongly disagree	8	1.2	3.20	0.70
	Disagree	84	12.8		
	Agree	330	50.3		
	Strongly agree	234	35.7		
I am attracted by the success of other entrepreneurs	Strongly disagree	14	2.1	3.23	0.72
	Disagree	68	10.4		
	Agree	324	49.4		
	Strongly agree	250	38.1		
I am encouraged by government's policies and incentives	Strongly disagree	30	4.6	2.80	0.75
	Disagree	173	26.4		
	Agree	349	53.2		
	Strongly agree	104	15.9		
I am encouraged by relatives and friends	Strongly disagree	29	4.4	2.74	0.75
	Disagree	201	30.6		
	Agree	335	51.1		
	Strongly agree	91	13.9		
I can sense the business opportunities	Strongly disagree	19	2.9	2.89	0.73
	Disagree	159	24.2		
	Agree	352	53.7		
	Strongly agree	126	19.2		

n=656

5 CONCLUSION

The Malaysian government has initiatives and provides many incentives and opportunities through its various agencies to boost entrepreneurial activities in the country (Aman et al., 2011). More and more emphasis in terms of providing courses and training within the universities' curriculum and co-curriculum activities have been organized and implemented. These continuous efforts and initiatives on the part of the Ministry of Higher Education have in many ways developed interest and confidence among students to consider venturing into business upon graduation. This is proven through this study, which found a high level of enthusiasm amongst students (70.4 %) to consider becoming an entrepreneur upon graduation and the fact that a huge majority (73 %) of the students agrees that the university has provided sufficient training and assistance with regard to the area of entrepreneurship.

REFERENCES

Aman, O., et al. (2011) "The success factors of food small medium entrepreneurs under the one district one industry programme: A case on food manufacturing entrepreneurs in Melaka." *Journal of Global Entrepreneurship*. Global Research Agency, 1(1), pp. 44–64.

Fatoki, O. O. (2010) "Graduate entrepreneurial intention in South Africa: Motivations and obstacles." *International Journal of Business and Management*. Canadian Center of Science and Education, 5(9), p. 87.

Fiet, J. O. (2001) "The theoretical side of teaching entrepreneurship." *Journal of Business Venturing*. Elsevier, 16(1), pp. 1–24.

GEM Consortium. (2017) *Global Report*. Global Entrepreneurship Research Association.

Gibb, A., and Ritchie, J. (1982) "Understanding the process of starting small businesses." *European Small Business Journal*. Sage Publications: Thousand Oaks, CA, 1(1), pp. 26–45.

Katundu, M. A., and Gabagambi, D. M. (2016) "Barriers to business start-up among Tanzanian university graduates: Evidence from the University of Dar-es-salaam." *Global Business Review*. SAGE Publications Sage India: New Delhi, India, 17(1), pp. 16–37.

Nowiński, W., and Haddoud, M. Y. (2019) "The role of inspiring role models in enhancing entrepreneurial intention." *Journal of Business Research*. Elsevier, 96, pp. 183–193.

Pekkala Kerr, S., Kerr, W., and Xu, T. (2017) *Personality Traits of Entrepreneurs: A Review of Recent Literature*. National Bureau of Economic Research, Inc.

Pulka, B. M., Rikwentishe, R., and Ibrahim, B. (2014) "An evaluation of students' attitude towards entrepreneurship education in some selected universities in North East Nigeria." *Global Journal of Management and Business Research*.

Facing Global Digital Revolution – Nirmala Arum Janie, Dwi Mulyaningsih & Wahyu Rachmawati (eds)
© 2020 Taylor & Francis Group, London, ISBN 978-0-367-33912-8

Conceptual framework of market orientation and firm performance toward the manufacturing sector: Moderating effect of innovation

S.W.M. Yusof, I.A. Majid & A.R. Yunus
FPTT, Universiti Teknikal Malaysia, Melaka, Malaysia

ABSTRACT: Market orientation constructs have been examined by a number of studies in marketing literature. In general, these studies demonstrate a favorable connection between market orientation, innovation, and firm performance. This article specifically examines market orientation in the context of manufacturing firms in Malaysia. A thorough examination of the existing literature is used to create a conceptual framework by investigating the main antecedents, the relationship between market orientation and performance, as well as innovation as moderator in that connection. This article further explores a number of studies of Malaysian manufacturing firms in different ways and provides recommendations for future research to gain an understanding of how market orientation affects manufacturing sector efficiency.

1 INTRODUCTION

According to Udriyah, Tham, and Azam (2019), market orientation, in theory, reflects an established idea using a broad range of approaches, techniques, and instruments, and can be regarded as a strategy for attaining viable competitive benefit. Broadly, market orientation is a culture in which elements such as customer orientation, competition orientation, interfunctional coordination, and responsiveness are highlighted as key to corporate achievement (Kohli & Jaworski, 1990; Narver & Slater, 1990). Although many studies have focused on market orientation–performance relationships, few have attempted to synthesize the current results in manufacturing businesses. With attention to MITI (Ministry of International Trade and Industry Malaysia, 2018), "Malaysia is anticipated to boost significantly up to 154 percent in the textiles, apparel and footwear industry." Previous studies demonstrate certain significant distinctions between bigger organizations and small-medium enterprises (SMEs) (Acs & Audretsch, 1987; Coviello, Brodie, & Munro, 2000), and the role of market orientations in bigger organizations can be expected to be very different from that of SMEs.

Empirical evidence, however, shows that in a developing country like Malaysia, the creation of risks is considered as a mechanism for improving revenue distribution, stimulating financial development, and redefining an economic framework that is extremely dependent on big companies' operations (Yusof et al., 2017). Therefore, a conceptual framework for the consideration of market orientations in manufacturing firms is the key goal of this article. This could help us better understand the role of market orientations and their impact on the performance of large organizations. Businesses can achieve efficiency by the use of market-oriented culture in the way they know market requirements, wants, and needs (Andiyanto, Miyasto, & Sufian, 2017). The relationships between market orientation and performance have received strong attention among researchers. This article explores various variables pertaining to market orientation, innovation, and business performance to sufficiently provide an acceptable basis in formulating a conceptual research framework.

2 THE CONCEPTUAL FRAMEWORK

Market orientation can be described as the company's technique of creating superior results as well as conduct required strategy to enhance the efficiency of the company, according to Maydeu-olivares and Lado (2003). Performance can be accomplished through the application of a market-oriented culture, which is how the company understands the market's needs, wants and demands (Andiyanto et al., 2017). Moreover, Kawano (2019) has pointed out that latecomers in manufacturing have to learn basic knowledge and technology before moving toward technological upgrading. Customer orientation, competitor orientation, and interfunctional cooperation are market orientation dimensions (Narver & Slater, 1990; Azam et al., 2014; Tham et al., 2017).

Marketing literature has traditionally seen market orientation as an important aspect of corporate culture. Market orientation's main antecedents can be seen in two classes, namely, structural variables that include behavioral elements of the organization as proposed by Narver and Slater (1990), and those established by Kohli and Jaworski (1990) as cultural variables that represent organizational member's standards and common values. Narver and Slater (1990) view market orientation in three aspects: client orientation, competitor orientation, and interfunctional coordination. Kohli and Jaworski (1990) define the broad dimension of market orientation as intelligence generation, intelligence distribution, and organization. Both methods have benefits and have been used by other scholars (Hult, Ketchen Jr, & Slater, 2005) before this integrative method.

Despite all the arguments and explanations illustrated in literature pertaining to the significance of market orientation as per Reniati (2013), innovation is an opportunity method that becomes a marketable concept. Manufacturing firms can profit even more if they create, interact, adopt, and explore innovation orientation (Saunila, 2014). Consequently, research has shown that innovation and market orientation have a beneficial connection to firm performance. For this reason, Aziz and Samad (2016) argue that manufacturing companies should use innovation to achieve a competitive advantage in the marketplace. Companies using innovative capacities and collective capacity are more likely to encourage efficiency, according to Amin et al. (2016), and it is suggested that manufacturing firms concentrate on refined innovations in order to boost their company efficiency.

According to Smith and Reece (1999), business performance is described as the operating capacity to fulfill the needs of significant shareholders of the company and is evaluated to assess the achievement of an organization. Business performance dimensions in this research are profitability, gross profit, return on investment, and sales growth. Ultimately, innovation and company performance have a substantial favorable connection (Najib & Kiminami, 2011; Saunila, 2014; Vazquez-avila., 2014; Herman, Hady, & Arafah, 2018).

The variables and some relevant discussions pertaining to market orientation are exhibited and summarized in Table 1.

Table 1. Summary of related literature pertaining to market orientation.

Author (Year)	Market orientation definitions	Variables and findings
Shapiro (1988)	Organization is market-oriented if, in any associated department, data on all-important purchasing factors will stream, making tactical and strategic determinations interdivisionally and interfunctionally. Well-coordinated and committed to decision-making in organizational departments.	Variables: Market Orientation Findings: A set of procedures that are relevant to all elements of the market-oriented organizations. It is better approach than just "getting close to the client."
Ruekert (1992)	Market orientation means that an organization acquires and uses client information; creates a system that addresses and	Variable: Market orientation from the view of organizational strategy and procedures, individual attitudes, and long-

(*Continued*)

Table 1. (*Continued*)

Author (Year)	Market orientation definitions	Variables and findings
	executes client problems by being responsive to client requirements and wants.	term economic performance. Findings: All executives of five SBUs (Strategic Business Units) try to create and retain a market orientation within the enterprise.
Deshpandé, Farley, and Webster (1993)	Market orientations are identical to customer orientation, which aims to create a company that benefits the client as well as other associates, such as executives and staff, for a long-term gain.	Variable: Customer orientation and business performance. Findings: The performance of the business was linked to the client rating of the customer orientation of the supplier.
Day (1994)	The market direction demonstrates the company's excellent ability to understand and satisfy customers.	Variables: Role of skills in generating organizations driven by the market. Findings: Excellent knowledge and customer satisfaction capacities can lead to excellent company results.
Kohli and Jaworski (1990)	The market orientation relates to the organization generating extensive market information linked to present and future requirements, disseminating intelligence across departments and organizations in response to produced and distributed intelligence.	Variables: Market Orientation Findings: The Market Orientation (MARKOR) scale of 32 points is used to engage in, disseminate, and react to produced intelligence in multi-department market intelligence operations.
Narver and Slater (1990)	The organizational culture, which can act to create the predominant values and performance for the organization on an ongoing basis.	Variables: Market orientation and business profitability. Findings: Market orientation and efficiency are closely linked to the management of important market and business impacts.

Based on the review of the related literature which explores the various relevant variables as forwarded by some prominent authors in the field, the following conceptual framework is proposed by the researcher.

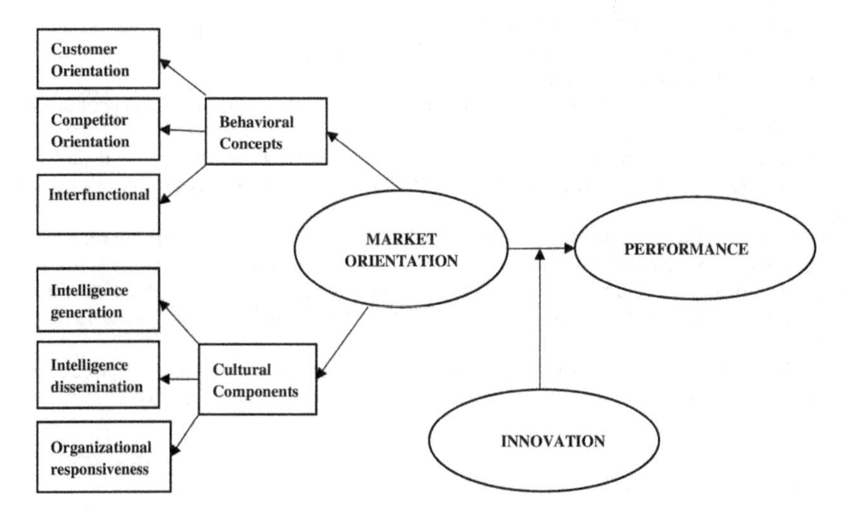

Figure 1. Conceptual framework of the moderating effect of innovation on the relationship between market orientation and performance.

3 CONCLUSION

The related literature has provided a strong understanding and evidence to support the formation of a research framework focusing on the moderating effect of innovation on the relationship between market orientation and performance. This study perceived market orientation to be comprised of behavioral concepts amongst the related stakeholders and cultural components of the organizations. The impact of market orientation and innovation on business performance may shed light to better understanding and offer some solutions to business organization.

REFERENCES

Amin, M. et al. (2016) "The effect of market orientation as a mediating variable in the relationship between entrepreneurial orientation and SMEs performance." *Nankai Business Review International*. Emerald Group Publishing Limited, 7(1), pp. 39–59.

Andiyanto, F., Miyasto, M., and Sufian, S. (2017) "Analisis pengaruh orientasi pasar dan inovasi terhadap keunggulan bersaing dalam rangka meningkatkan kinerja bisnis." Diponegoro University, pp. 1–20.

Azam, S. M. F. et al. (2014) "Training Program Effectiveness of Service Initiators: Measuring Perception of Female Employees of bank using logistic approach." *Asian Research Journal of Business Management*, 1(2), pp. 98–108.

Aziz, N. N. A., and Samad, S. (2016) "Innovation and competitive advantage: Moderating effects of firm age in foods manufacturing SMEs in Malaysia." *Procedia Economics and Finance*. Elsevier, 35, pp. 256–266.

Coviello, N. E., Brodie, R. J., and Munro, H. J. (2000) "An investigation of marketing practice by firm size." *Journal of Business Venturing*. Elsevier, 15(5–6), pp. 523–545.

Herman, H., Hady, H., and Arafah, W. (2018) "The influence of market orientation and product innovation on the competitive advantage and its implication toward Small and Medium Enterprises (UKM) performance." *International Journal of Science and Engineering Invention*, 4(8).

Hult, G. T. M., Ketchen Jr, D. J., and Slater, S. F. (2005) "Market orientation and performance: An integration of disparate approaches." *Strategic Management Journal*. Wiley Online Library, 26(12), pp. 1173–1181.

Kawano, M. (2019) "Changing resource-based manufacturing industry: The case of the rubber industry in Malaysia and Thailand." in *Emerging States at Crossroads*. Springer, pp. 145–162.

Kohli, A. K., and Jaworski, B. J. (1990) "Market orientation: The construct, research propositions, and managerial implications." *Journal of marketing*. Sage Publications: Los Angeles, CA, 54(2), pp. 1–18.

Ministry of International Trade and Industry Malaysia (2018) *Textile, Apparel footwear industry profile*, http://www.miti.gov.my/.

Najib, M., and Kiminami, A. (2011) "Innovation, cooperation and business performance: Some evidence from Indonesian small food processing cluster." *Journal of Agribusiness in Developing and Emerging Economies*. Emerald Group Publishing Limited, 1(1), pp. 75–96.

Narver, J. C., and Slater, S. F. (1990) "The effect of a market orientation on business profitability." *Journal of Marketing*. Sage Publications: Los Angeles, CA, 54(4), pp. 20–35.

Saunila, M. (2014) "Innovation capability for SME success: Perspectives of financial and operational performance." *Journal of Advances in Management Research*. Emerald Group Publishing Limited, 11 (2), pp. 163–175.

Tham, J. et al. (2017) "Internet and data security: Understanding customer perception on trusting virtual banking security in Malaysia." *European Journal of Social Sciences Studies*.

Udriyah, U., Tham, J., and Azam, S. (2019) "The effects of market orientation and innovation on competitive advantage and business performance of textile SMEs." *Management Science Letters*, 9(9), pp. 1419–1428.

Yusof, S. W. M. et al. (2017) "Exploring the cultural determinants of entrepreneurial success: The case of Malaysia." *International Journal of Advanced and Applied Sciences*, 4(12), pp. 287–297.

Facing Global Digital Revolution – Nirmala Arum Janie, Dwi Mulyaningsih & Wahyu Rachmawati (eds)
 ISBN 978-0-367-33912-8

The essence of information governance for public sector auditing in Malaysia

R.M. Ali
The National Library of Malaysia, Kuala Lumpur, Malaysia

N.M. Nordin
Universiti Teknologi MARA (UiTM), Machang, Malaysia

S.M. Sharif
Universiti Teknikal Malaysia Melaka (UTeM), Melaka, Malaysia

A.M. Isa
Universiti Teknologi MARA (UiTM), Machang, Malaysia

ABSTRACT: There is an increasing demand for accountability and transparency in the public sector in Malaysia, which has resulted in a change of the political landscape. The public sector is the backbone of effective sustainable development, whilst good governance is a prerequisite. There must be a checks-and-balances mechanism through public auditing that ensures national administration and development are consistent with rules and regulations. To this end, public auditors play a crucial role in verifying the implementation of those activities. Due to the fact that mismanagement, corruption, and fraud are rampant today, information governance would be the answer to curb the crimes, by providing a set of processes, roles, policies, standards, and measures, which together ensure the efficient use of information and thus the achievement of an organization's goals. This research adopts simple random sampling design involving interviews with six senior public sector auditors. The interviews were transcribed and analyzed using NVivo11 statistical analysis tool. It is found that information is the core of public auditing. Findings proved that shortages of and incomplete records impede effective public auditing as auditors spend considerable time looking for relevant records. Information governance is imperative if public auditing is to be more effective, trusted, and respected by members of the public.

Keywords: accountability, auditing, information governance, records management, public sector

1 INTRODUCTION

There is an increasing demand for accountability and transparency in the public sector in Malaysia, which partly has resulted in a change of the political landscape. The public sector is the backbone of effective sustainable development, whilst good governance is a prerequisite. There must be a checks-and-balances mechanism through public auditing that ensures national administration and development are consistent with rules and regulations. To this end, public auditors play a crucial role in verifying the implementation of those activities is in accordance with standards and legal requirements. Records are by-products of business transactions that provide evidence of the activities. They are crucial to safeguard or, ironically, to implicate individuals pertaining to their role in a business transaction. Thus, ensuring the availability of trustworthy and timely records is essential in underpinning effective public audits. Information governance would be the best answer as it provides a holistic approach to managing records and information from creation to disposal. Therefore, it is imperative to investigate information governance–related problems encountered by public auditors in performing their tasks. The findings of the research would provide an insight on the implementation of information governance in the public sector.

2 PUBLIC AUDITING

To date, the role of public auditing has become more crucial than ever as the Auditor General's Audit Report is a much-awaited document by members of the public, particularly politicians. This is to say that the National Audit Department (NAD) and public sector auditors play a critical role in ensuring all authorized activities are done economically, effectively, and efficiently (National Audit Department, 2017) and public money is being properly utilized, safeguarded, and recorded. This is in line with the mandate given to the Auditor General (AG) under the provisions of the Federal Constitution and the Audit Act (1957) Section 6. Their main focus is monitoring and evaluating government activities to see whether programs or projects have been carried out efficiently and economically in order to achieve desired objectives and goals. Hence, the role of public sector auditors has constantly evolved due to exposure and concern for better governance and greater accountability in managing public funds. There are four types of audits that are usually conducted: compliance audits, financial audits, special audits, and performance audits and these cover the areas of construction, law enforcement, asset management, socioeconomic programs, ICT, service procurement, and privatization. One of the most crucial audits is the financial audit that ensures the quality of financial management.

3 WHY INFORMATION GOVERNANCE?

Information is the main factor empowering our society and economy today. The growth of information has been exponential since the existence of the Internet and social media. This has led to a more challenging situation where misuse, leakage, and false information contribute to the complexity (Sharif et al., 2018). To this end, information governance enables organizations to take advantage of the benefits of improving technology and prepare to adapt to any changes ahead. Organizations will gain public trust when they are transparent, although transparency alone is inadequate to satisfy public trust (Sharif, 2009). Hence, it is essential for each employee to understand the key attributes of good governance such as accountability, responsibility, responsiveness, and participation. On the other hand, the absence of transparency and accountability will lead the organization to loss of reputation, lawsuits, and fraud charges (Isa & Nordin, 2013).

Information governance is a comprehensive approach to a different instrument that requires a secured information exchange. It aims to maximize the value of information to the organization and protect information within its own life cycle (Kooper, Maes, & Lindgreen, 2011). Hence, information governance will be able to deliver information security, information quality, and metadata domains (Khatri & Brown, 2010). Indeed, the main focus of information governance is the transformation of information from liability to strategic asset (Smith, 2015). Information governance enhances the trustworthiness of information if supported by organizational strategy and commitment particularly from top management. This ensures that accurate information is available at the right time and at the right place. To this end, complete records will be one of the cornerstones of effective management in an organization. In today's environment, managing information and records is an increasingly important preoccupation of information governance. Keeping a genuine, accurate, and complete record is one of the vital elements of effective information governance (Isa et al., 2015). Information governance is an organization's framework in managing information during its life cycle and in supporting the organization's policy, strategy, regulatory, legal, risk, and environmental requirements (AHIMA, 2017). It is evident that trustworthy, accurate, and timely information enables an organization to achieve its mission and goals.

4 RESEARCH OBJECTIVE AND METHODOLOGY

The main objective of the research is to identify the most common information governance–related problems encountered by public auditors. The research adopts qualitative methodology using a case study at the NAD, Malaysia. The population of the study was public auditors. Respondents were chosen through purposive sampling design involving six public sector auditors with auditing experience ranging between six to more than twenty years. Data were collected through semi-structured interviews that were subsequently transcribed and analyzed using NVivo11 statistical analysis tool.

5 FINDINGS AND DISCUSSION

Findings in Figure 1 proved that the auditors were aware of the significant role of information facilitating them to achieve accountability and integrity that would enhance public confidence in the audit. Internally, the NAD has its own policy for managing information in tandem with the Official Secrets Act 1972 (Act 88), which requires records and information to be classified into categories, namely, top secret, secret, confidential and restricted. In practical terms, public sector auditors take an excessive amount of time to locate evidence or records they need if the records are not properly kept. Simply, if they do not have supporting documents, the audit process will fail. Therefore, they could not analyze or give their opinion or recommendations in their audit report. Hence, this would create a challenge to the auditors and, as a result, the auditors'.credibility would be at risk.

"Governance" is the second most frequent word used by the respondents. They believe that it is crucial for the government to give high attention to information compliance, particularly in terms of security and privacy. This enables the public sector to establish guidelines on how to govern the information accurately and with integrity.Auditors deal with highly confidential information and its trustworthiness and accuracy must remain intact over time. This can only be ascertained through information governance that comprises policies and ethical values among public servants. The respondents also believe that such policies would impede information leakage, which can easily be done with sophisticated smart phones and other gadgets. Information technology teams cannot be expected to handle all the information security issues alone (Iannarelli & O'Shaughnessy, 2015) in the absence of a comprehensive policy. Thus, the implementation of information governance is imperative.

Figure 1. Word cloud of high frequency words by all respondents.

The respondents also addressed the absence of clearly designated responsibility in managing records that has led to incomplete records, and a shortage of records. Effective record keeping can only be implemented when every employee plays their role in ensuring records are completely captured from the creation stage through to their legal disposal, so that the availability of trustworthy, complete, and timely records is certain. This situation can only be changed with the presence of commitment and involvement of senior management. It has to be a top-down approach for the impact to be immediate. The findings proved that the implementation of information governance is imperative.

6 CONCLUSION AND FUTURE RESEARCH

Public auditing plays a critical role in ensuring all authorized activities are done economically, effectively, and efficiently. However, this can only be achieved with the availability of trustworthy and timely records. Unfortunately, shortage of and incomplete records impede effective public auditing as auditors spend considerable time looking for relevant records, which sometimes proves futile. This situation cannot be allowed to continue if the government is serious about improving the performance of public departments and at the same becoming more transparent. Future research should investigate the actual situation in any public department in order to determine underlying issues that lead to poor information governance. Arguably, the need for information governance is imperative as it facilitates effective public auditing, which in turn proves that the government and its machineries have utilized and safeguard public money in line with procedures and regulations for the administration and development of the nation.

ACKNOWLEDGMENT

The authors would like to thank the NAD for their support and cooperation in accomplishing this research.

REFERENCES

AHIMA (2017) *American Health Information Management Association (AHIMA)*, http://www.ahima.org/.

Iannarelli, J. G., and O'Shaughnessy, M. (2015) "The changing landscape." *Information Governance and Security*, pp. 45–53.

Isa, A. M., et al. (2015) "Information seeking behavior among IT professionals in using electronic records management system." *Advanced Science Letters*. American Scientific Publishers, 21(6), pp. 2105–2108.

Isa, A., and Nordin, N. (2013) "Strategic records and risk management for the sustainability of organizations: A case study investigation." *Comma*. Liverpool University Press, 2013(1), pp. 29–40.

Khatri, V., and Brown, C. V (2010) "Designing data governance." *Communications of the ACM*. ACM, 53(1), pp. 148–152.

Kooper, M. N., Maes, R., and Lindgreen, E. E. O. R. (2011) "On the governance of information: Introducing a new concept of governance to support the management of information." *International Journal of Information Management*. Elsevier, 31(3), pp. 195–200.

National Audit Department (2017) *Audit Acts 1975*, http://www.audit.gov.my.

Sharif, S., et al. (2018) "Role of values and competencies in university intellectual property commercialization: A critical review." *Turkish Online Journal of Design Art And Communication*, 8, pp. 887–904.

Sharif, S. M. (2009) *Generasi Ulul Albab: Segunung Harapan Seteguh Gagasan*. Karisma.

Smith, M. S. (2015) "A phenomenological study of critical success factors in implementing information governance." University of Phoenix.

Facing Global Digital Revolution – Nirmala Arum Janie,
Dwi Mulyaningsih & Wahyu Rachmawati (eds)
© 2020 Taylor & Francis Group, London, ISBN 978-0-367-33912-8

A critique on the use of family ownership on accounting research: The Indonesia sociopolitical context

A.S. Wahyuni
Politeknik Negeri Ujung Pandang, Makassar, Indonesia

A. Chariri
Universitas Diponegoro, Semarang, Indonesia

ABSTRACT: This study is intended to critically analyze the views of Indonesian accounting researchers who employ family ownership as a variable in capital market-based accounting research. A qualitative approach is used in this study by borrowing literary critical theory, with the object of research: Sanjaya's article. This article was presented in the 17th Indonesia National Symposium on Accounting. Findings of this study showed that Sanjaya failed to recognize the condition of Indonesian society – in which the patterns of sociopolitical thought are formed by the "Orde Baru" regime – when he uses family ownership as a variable. Thus, agency theory, which is widely employed in accounting studies, is believed irrelevant to the Indonesian sociopolitical environment if the concept is not modified to the Indonesia setting.

1 INTRODUCTION

Accounting research, especially capital market-based accounting research, has dominated academic publication in a number of prestigious journals around the world. At least two seminal papers have been referred as the main papers in any publication related to such accounting research: Beaver (1968) and Ball and Brown (1968). These two papers were also cited by Indonesian accounting researchers including Sanjaya (2014) in his paper entitled "Keluarga sebagai Pemilik Ultimat dan Kinerja Perusahaan" (Family as the ultimate owner and firm performance), presented at the 17th National Symposium on Accounting (SNA) – a prestigious national convention for Indonesian accounting scholars.

If we refer to Sanjaya's paper, it can be seen that his research objective is to investigate whether family ownership as controlling shareholder affects firm performance. Using 604 samples of manufacturing companies listed in the Indonesia Stock Exchanges, this study used ultimate ownership to identify family ownership or non-family ownership. The research findings showed that family ownership negatively affected firm performance. This inferred that family ownership tends to decrease firm performance. The study suggested that entrenchment effect is more dominant than alignment effect on family ownership in Indonesia.

Sanjaya's paper may contribute interesting findings in regard to the effect of family ownership on firm performance. However, he ignored an important issue: namely, that the characteristics of family ownership in Indonesia are different from those in Western countries. Indeed, Sanjaya ignored the Indonesian sociopolitical environment that was internalized by the "Orde Baru" (New Order) regime with President Soeharto as a leader.

Sanjaya's paper was selected as research object because the data used by Sanjaya (2014) were data available after the Orde Baru era. This makes it relevant to reconcile with the study by Shiraishi (2009), which analyzed the influence of sociopolitical rulers formed by the Orde Baru regime on the establishment of a family. Shiraishi's study can be further used to analyze the context of "family" in Indonesia, in which the function of "family" is questionable in the light of family ownership in the Indonesian publicly listed companies. Although the Orde Baru regime collapsed in 1998, the ideology of Orde Baru is still internalized in the minds of Indonesian people, including the business community. Yet here lies the importance of this research: social and political conditions have to be considered by accounting scholars in capital market-based accounting research, especially when they use family ownership as a variable.

2 RESEARCH METHOD

The use of a single research paper to analyze the conditions affecting the views of researchers and how they understand various external influences is common in literary areas. However, it is not easy to find the use of literary criticism in accounting research areas, especially market-based accounting research in Indonesia. Literary criticism has been used in the accounting discipline, for example, Schuler (1998). But this study combined symbolic interpretation and literary criticism, which is actually not literary criticism. Literary criticism is an attempt to analyze text and relate it to the situation in which the work (text) is created and published.

Sanjaya's paper (2014) was chosen because it is seen as the most current study on family ownership and the relevance of the employed data. Finally, Sanjaya's views in regard to the construction of family ownership as a variable will be understood by other researchers in Indonesia in general. In this case, Sanjaya's paper is used because his ideas represent the ideas of others in regard to the use of family ownership in the Indonesia context. For the purposes of writing this research, the authors use a sociopolitical approach. This is because such an approach is relevant to Shiraishi's study (2009) on the establishment of the image of Indonesian family to assess the social and political conditions at the Soeharto era.

3 RESEARCH FINDINGS AND DISCUSSIONS

3.1 *Family in the Indonesian sociopolitical context*

Shiraishi (2009), a Japanese ethnographer, has been looking for the origin of the formation of the word "keluarga (family)" in Indonesia. Family is perceived as basis of creating nationality. Each nation, according to Shiraishi, has its own custom and style in establishing its nationality. The image of a nation always changes in line with the dynamics of its history. Moreover, Freire (2008) asserted that, in general, the parent–child relationship in a family reflects the objective conditions of the culture of the surrounding social structure. Hence, the relations among classes in a society can be seen from the family relationships in that society.

To understand the Indonesian family comprehensively, as Shiraishi (2009: 7) claims, any discourses in society regarding family must be carefully assessed to avoid taking for granted beliefs claiming that family is only concerned with father, mother, and children. A description of how families are formed in the Orde Baru era can be seen from the Soeharto's statement as written by Shiraishi (2009: 1):

> *"In my eyes there is neither gold nor a child that I do not like. Nothing at all. All of them, in their tasks and respective fields, get the same trust from me. All my closed servants [ministers] are fit with their fields . . . "* (translation)

It is clear that Soeharto put himself as a father and his servants (ministers) are seen as his children. During his 32 years in power, Soeharto put himself as a Supreme Father (the God Father) for his subordinates and people. Children must obey their father/mother and they are not allowed to criticize father/mother. Any activities against or opposed him is considered as reprehensible, even if the children know that the father's actions or behavior is not in accordance with ethical and moral behavior; for example, illegally taking public money.

The Orde Baru regime then established the nation's ruler to be like a harmonious family. By doing so, it is impossible for children to act against their father. Shiraishi (2009: 122) claims that

> *"...in spite of his own thoughts and opinions, "children" are expected to be mature enough to control themselves and continue to adhere to the instructions and orders from Soeharto. In exchanges, the father/mother will provide their children with protection and security."* (translation)

The protection provided by is eventually transformed into nepotism. Those who were close to Soeharto got special access. This can be seen from a number of companies that are backed by the Cendana family and its cronies. Furthermore, what happens to the "national family" since the establishment of the Orde Baru era and what the Indonesia people do to the tradition of "national family" are intertwined. Attitudes and behaviors among children and their father become a pattern in a wider institution, called as Indonesia.

The Orde Baru pattern of family is then copied by his subordinates, both in the level of provincial government and municipality as well as other institutions, including in education and profit-oriented companies. Indeed, companies that are actually not state-owned internalize their beliefs in terms of the father–child relationship to create order and security for their business activities. The following statement by Ruth McVey as quoted by Shiraishi (2009: 122) clearly describes how the concept of family is internalized to various institutions in Indonesia.

> *"The organization of a group or government and society in term of "basic family" will become continuous themes [emphasis added] in the study of Indonesia modern. It is used to justify paternalism-often ridiculed as bapakisme [father]- in government or social relationships [emphasis added] and to prevent the internal challenges against leadership ... "* (translation)

Since the Orde Baru era, companies in Indonesia not only implement a father–son relationship but also create a connection with "the father" of the origin country directly. Some of them, as claimed by Shiraishi, are Bank Duta, PT Citra Sari Makmur, and PT Sumber Tani Agung. Such internalization has affected the interaction with all stakeholders in the company, especially if the company is characterized by family ownership (usually majority shareholding).

3.2 *SNA and representation of accounting scholars*

SNA is a scholarly activity – a national agenda of one of the largest professional institutions, namely the Indonesian Institute of Accountants (IAI) – which is conducted as a communication medium among Indonesian accounting scholars to share accounting research findings and to build networks. SNA represents the development of accounting studies in the Indonesia context from year to year. Thus, any accounting research papers accepted for presentation in any SNA can be considered as the best accounting research papers in Indonesia.

SNA has high bargaining power, considering the increasing numbers of participants attending the event. The statement by Professor Zaki Baridwan – the former Chairman of the Accounting Academician Compartment – may reflect the importance of SNA. He claimed that at the beginning of SNA events, its participants were no more than 50 accounting scholars. However, by 2013, the number of participants joining the event had increased dramatically and there were 176 accepted papers 176, excluding the papers rejected by the Committee. In the last five years, the number of accepted papers is much lower than in the previous years. An interesting justification can be seen from the statement of accounting scholars from a private university in Malang, East Java, who posted his comment on social media, 25 days before the 2014 SNA agenda:

"Banglore-India,... this manuscript is the same one that I submitted to the SNA Lombok (only different language), and it was rejected [by the SNA Committee] This means that SNA is much more difficult than AAAA [Asian Academic Accounting Association]..."

The statement shows that the SNA committee decided on the accepted manuscripts carefully. The manuscript was rejected by the SNA Committee, but at an international conference the same manuscript was accepted for presentation. Thus, manuscripts accepted in the SNA represent better quality than the international standards. In other words, all accepted manuscripts of SNA are the best ones, including all manuscripts categorized in the capital market-based accounting research themes. This is the reason why we chose manuscripts concerning the use of family ownership in the capital market-based accounting research as the object of our analysis.

3.3 *Family ownership and agency theory*

Considering the characteristics of ownership, the Indonesian scholars who use family ownership as a research variable are trapped in the agency theory initiated by Jensen and Meckling (1976). This can be seen from a number of articles accepted for presentation at the 2014 SNA (for example, Siagian, 2014; Brian and Martani, 2014; Nuritomo, 2014; and Ridha, 2014). The articles plainly used agency theory as the basis of forming a hypothesis, regardless of the different characteristics of family ownership of the Indonesian firms. Indeed, two articles (Putri, 2014, and Ulupui, 2014) enjoyed citing Jensen and Meckling (1976) arguments and wrote them in the first paragraph of the introduction section. In her article, Putri (2014) wrote: "According to the agency theory ... the agency conflict depends on the patterns of the company ownership." Furthermore, Ulupui (2014) in his article claims, "The compensation distributed to the executives is an interesting phenomenon to study." The history of modern research on compensation began in the early 1980s, that is the emergence of agency theory by Berle and Means (1932), which was subsequently formalized by Jensen and Meckling (1976) (quoted by Murphy, 1998). These two quotations imply that the ownership characteristics among capital markets are assumed to be the same.

The accounting scholars have ignored the fact that the ownership structure of the Indonesian publicly listed companies is concentrated. Hapsoro's (2007) study insisted that the ownership structure of Indonesia companies is different from that in the United States and United Kingdom. Indeed, most Indonesian companies are concentrated and affiliated with the same family. Research by Lim (2012), also confirmed that companies in the media industry are concentrated after the Soeharto era in 1998. Another empirical finding shows that 70% of publicly listed companies in Indonesia are controlled by their own family members (Claessens, Djankov, & Lang, 2000). Fourteen years later, Nuritomo (2014) found interesting evidence that public ownership of Indonesian companies had decreased since 2004. This implies that the ownership structure of publicly listed companies had become more concentrated.

Using family ownership as a variable without considering its context could lead to bias findings. Sanjaya (2014: 14) for example incorporated agency theory in his article and explicitly claimed:

"In the case of dispersed ownership of publicly listed companies, the power of supervision by the owner becomes disappeared. Managers are more interested in thinking about their own interests rather than that of shareholders. This leads to a conflict of interest between shareholders and managers well known as the principal-agent' theory."

Similar views can be seen from other Indonesian accounting scholars, for example, articles written by Christian and Mustamu (2013). Unfortunately, the scholars failed to analyze how family relationship influences patterns of ownership, which then constructs the unique characteristics of family ownership in Indonesia. Thus, ignoring family ownership issues will lead to questionable findings on all studies considering family ownership as a main variable.

4 CONCLUSION

The new order under the reins of Soeharto leadership was the formative years of anti-criticism mindsets. The status quo made Soeharto comfortable occupying the presidential chair, and then drain the desire to keep maintaining it as the holder of the highest state authority. For the sake of perpetuating power, Soeharto then made himself a "father" who protected the country and as such must be respected and obeyed by "children," all Indonesian people. This is exactly what Sanjaya (2014) ignored in his article. Sanjaya as one of the researchers interested in the capital market-based accounting studies, snubbed the patterns of Indonesian thought formed by the system during the new order. Not only Sanjaya; other accounting scholars with similar research interests using family ownership as a variable also ignored it. Father-son relationships inherited from Soeharto as a symbol of obedience and submission to the Supreme Father had flowed and crop up at various institutions in Indonesia, including publicly listed companies. Ultimately, agency theory may become irrelevant in accounting studies if not modified to account for the Indonesia context or not arranged with clear control variables.

Departing from our critical analysis, we recommend that further capital market studies using family ownership should consider the aspect of "child" compliance that is symbolized by the family who manage publicly listed companies with "father" as a symbol of family shareholders (the controlling shareholders). In other words, when we use family ownership, we should not consider family relationships based on marital status but need to consider how close someone's relationship is with the controlling shareholders. By considering this, we believe that future studies using family ownership will produce more accurate and valid findings, reflecting the Indonesian context as a country with unique history and social and political conditions.

REFERENCES

Ball, R., and Brown, P. (1968) "An Empirical Evaluation of Accounting Income Numbers." *Journal of Accounting Reserach*, 6, pp. 159–177.

Bandel, K. (2013) "Sastra Nasionalisme Pascakolonialitas." *Yogyakarta*: *Pustaka Hariara*.

Beaver, W. (1968) "The information content of annual earnings announcements." *Journal Accounting Research*, Supplement, 6, pp. 67–92.

Brian, I., and Martani, D. (2014) "Analisis Pengaruh Penghindaran Pajak dan Kepemilikan Keluarga terhadap Waktu Pengumuman Laporan Kuangan Tahunan Perusahaan." *Simposium Nasional Akuntansi*, 17.

Brundy, E. P., and Siswantaya, I. G. (2014) "Pengaruh Mekanisme Pengawasan terhadap Aktivitas "Tunneling." *Simposium Nasional Akuntansi* 17.

Christian, B., and Mustamu, R. H. (2013) "Penerapan Prinsip-prinsip Good Corporate Governance: Studi Deskriptif pada Perusahaan Keluarga Jasa Properti." *Jurnal Agora*, 1(3).

Claessens, S., Djankov, S., and Lang, L.H.P. (2000) "The separation of ownership and control in East Asian corporations." *Journal of Financial Economics*, 58(1/2), pp. 81–112.

Freire, P. (2008) "Pendidikan Kaum Tertindas." *Jakarta: Pustaka LP3ES Indonesia*.

Hapsoro, D. (2007) "Pengaruh Struktur Kepemilikan Terhadap Transparasi: Studi Empiris di Pasar Modal Indonesia." *Jurnal Akuntansi & Manajemen*, 18(2), pp. 65–85.

Jensen, M. C., and Meckling, W. H. (1976) "Theory of the firm: Managerial behavior, agency costs, and ownership structure." *Journal of Financial Economics*, 3, pp. 305–360.

Lim, M. (2012) "The league of thirteen: Media concentration in Indonesia: Research report." Tempe, AZ: Participatory Media Lab at Arizona State University.

Nuritomo and Martani, D. (2014) "Insentif Pajak, Kepemilikan, dan Penghindaran Pajak Perusahaan: Studi Penerapan Peraturan Pemerintah No. 81 Tahun 2007." *Simposium Nasional Akuntansi* 17.

Putri, M., and Diyanty, V. (2014) "Pengaruh Struktur Kepemilikan dan Mekanisme Corporate Governance terhadap Tingkat Pengungkapan Laporan Keberlanjutan." *Simposium Nasional Akuntansi* 17.

Ridha, M., and Martani. D. (2014) "Analisis terhadap Agresivitas Pajak, Agresivitas Pelaporan Keuangan, Kepemilikan Keluarga, dan Tata Kelola Perusahaan di Indonesia." *Simposium Nasional Akuntansi* 17.

Sanjaya, I. P. S. (2014) "Keluarga sebagai Pemilik Ultimat dan Kinerja Perusahaan." *Simposium Nasional Akuntansi* 17.

Schuler, D. K. (1998) "Redefining a certified public accounting firm." *Accounting, Organizations and Society*, 3(5/6), pp. 549–567.

Shiraishi, S. S. (2009) "Pahlawan pahlawan Belia: Keluarga Indonesia dalam Politik." *Jakarta: Nalar.*

Siagian, O. I., and Martani, D. (2014) "Analisis Pengaruh Perubahan Liabilitas Pajak Tangguhan Bersih, Tata Kelola Perusahaan, dan Kepemilikan Keluarga terhadap Manajemen Laba di Indonesia." *Simposium Nasional Akuntansi* 17.

Ulupui, I. G. K. A., Utama, S., and Karnen. K.A. (2014) "Pengaruh Kepemilikan Keluarga, Kedekatan Direksi & Komisaris dengan Pemilik Pengendali terhadap Kompensasi Direksi & Komisaris Perusahaan di Pasar Modal Indonesia." *Simposium Nasional Akuntansi* 17.

Facing Global Digital Revolution – Nirmala Arum Janie, Dwi Mulyaningsih & Wahyu Rachmawati (eds)
© 2020 Taylor & Francis Group, London, ISBN 978-0-367-33912-8

The impact of POKDAKAN group level on business innovations of small fish producers in Central Java

Albert
Universitas Semarang, Semarang, Indonesia

T. Elfitasari
Universitas Diponegoro, Semarang, Indonesia

ABSTRACT: This research aims to analyze whether different POKDAKAN level has an impact on implementing business innovations by fish farmers and to analyze if different POKDAKAN levels have different business innovations preferences. Research method used a quantitative approach involving distribution of questionnaires to fish farmers from different POKDAKAN levels. Results showed that fish farmers from different POKDAKAN levels showed significant differences in implementing business innovations. There were three business innovations that were found to differ significantly when carried out by small-scale fish farmers who belonged to different levels of fish farmer groups: seed production, market extension, and fish species extension. Different levels of POKDAKAN also seem to have different interest in business innovations preferences. Fish farmers from advanced levels are more interested in implementing market extension and fish species extension, while starter groups prefer to produce their own fish seed.

1 INTRODUCTION

The aquaculture industry has expanded, diversified, and advanced technologically, and has become the fastest food production sector in the world (Jiansan et al., 2001). This statement strongly supports the view that aquaculture is now becoming the number one source of fish supply globally and that it is continuously developing. It shows that the prospects for aquaculture are bright, especially for developing countries in the Asian region. Likewise, Central Java is also experiencing an increase with huge potential, with land utilization at 16% with plenty more still available (Central Java Statistical Center, 2016). Fish farming production in Central Java increased by 7.49% per year, in the report year 2002–2004 (Central Java Fisheries and Marine Affairs Office, 2005). Human capital also contributes to supporting the aquaculture industry in Indonesia. In 2014, Central Java's population reached more than 33.5 million people. Therefore, there is enormous potential for further aquaculture development in Central Java. This research aims to determine whether the level of POKDAKAN influenced the implementation of business innovations. The aim was also to determine whether the level influenced the POKDAKANs' preferences for the types of business innovations implemented.

1.1 *Business innovations of fish farmers*

Carland et al. (2007) identified five business innovation characteristics that differentiate entrepreneurial ventures from small businesses: innovating new product, innovating new means of manufacturing, expanding market distribution, expanding supplier resources, and reorganizing the industry. Based on the two characteristics of taking innovative action and seeking new markets, this research elaborated six fish farming business innovations. Five business innovations had been developed as innovative actions: fish feed production, fish seed production, fish species extension, product range extension, and utilization of other fish parts. The sixth entrepreneurial activity, market extension, was developed from the second entrepreneurial characteristics identified by Carland et al. (2007), namely seeking a new market.

1.2 *Fish farmer groups (POKDAKAN)*

Semarang's Marine Affairs and Fisheries Office have publicly announced their support for aquaculture development in Indonesia. They pointed out that the efficiency and productivity of the individual fish farmer depended on the capacity of the fish farmer to develop his or her own aquaculture business. In order to strengthen the fish farmer and family's position as part of the developing community, the Indonesian government has recommended that fish farmers establish and merge into POKDAKAN groups (Ministry of Marine Affairs and Fisheries, 2006).

In general, there are four accreditation levels assigned to POKDAKAN (Semarang Marine Affairs and Fisheries Office, 2005). In the following section the group levels are given in the original Indonesian language, and, because there are no direct translations for these descriptions into English, are identified as Level 1 to Level 4.

1. Kelas Pemula (Level 1 – the lowest level or starter/novice level). It is the lowest level, and usually, newly formed POKDAKANs are assigned to this level. They have the lowest capabilities according to the ten evaluation criteria. The village head signs the certificate for this level.
2. Kelas Lanjut (Level 2 – the secondary level). It is a step up from Level 1. Evaluation of the POKDAKAN members shows that they have a higher level of abilities according to the ten evaluation criteria than Level 1 POKDAKAN. The Head of District officer, in Indonesia known as Camat, signs the certificate for this level.
3. Kelas Madya (Level 3 – the mid advance level). The POKDAKAN at this level has much higher levels of abilities than the POKDAKANs at Levels 1 or 2, as measured according to the ten evaluation criteria. The Head of the Region officer, in Indonesia known as Walikota or Bupati, signs the certificate for this level.
4. Kelas Utama (Level 4 – the advanced level). It is the highest level of POKDAKAN achievement. The members perform at the highest level according to all ten evaluation criteria. The Governor of the province where the POKDAKAN is located signs the certificate for this level.

Many of the fish farmers believe that the POKDAKAN not only benefits their livelihoods but that they receive additional benefits from membership. The benefits that fish farmers have identified are income growth (Kadin Sumbar, 2008), soft loans from the bank, fish disease handling support (Suara Merdeka, 2002), government subsidies (Antaranews, 2007; Suara Merdeka, 2007), and many more.

2 METHODOLOGY

This research used a quantitative research approach, involving the distribution of questionnaires. Respondents were purposively chosen to avoid an imbalance between fish farmers from the starter and advanced POKDAKANs level. There are approximately 400 fish farmers from different levels and different regions in Central Java.

Data collected were then analyzed statistically using SPSS. For the simplification of the groupings for the analysis, the fish farmer group class are categorized into two: (i) starter/novice category (which included the starter and second level classes), and (ii) advanced category (which included the third and fourth level classes) (Pallant, 2016).

3 RESULT AND DISCUSSION

The analysis showed that the different fish farmer group levels resulted in significant differences between the implementation of the following business innovations: fish seed production, market extension, and fish species extension. Fish feed production and utilization of other fish parts also show significant p-values; however, since there were no participants in one of these categories for these two activities, the significant effect was not valid.

Since the majority of respondents in this research belonged to an advanced POKDAKAN, most respondents were members of either a level three or a level four POKDAKAN. According to the evaluation criteria, an advanced POKDAKAN should have a higher ability to manage and develop their business than a starter POKDAKAN. Their ability to manage and develop their business may include the ability to carry out business innovations since the evaluation criteria include the following performance components: (a) the ability to search out and use information and (b) the ability to plan activities which can increase their productivity. Therefore, evaluation of the entrepreneurial drive is made through the identification of capability to search out and use any information obtained to increase their productivity.

This study has identified the following four different aspects of business innovations between POKDAKAN levels: fish seed production, market extension, fish species extension, and utilization of other fish parts. The business innovations that did not show any significant differences between POKDAKAN levels were fish feed production and product range extension.

Table 1 also supports the expected outcome that fish farmers who belonged to advanced POKDAKANS were more likely to implement business innovations than fish farmers who were members of starter-level or novice POKDAKANS. Advanced POKDAKAN fish farmers (in contrast to starter POKDAKANS/novices) perform the following activities: market extension, fish species extension, product range extension, and utilization of other fish parts. This finding matches the expectation that fish farmers from higher-level POKDAKANS implement more business innovations than their starter-group counterparts.

Table 1. POKDAKAN groups (starter and advanced) towards business innovations.

Business innovations	p	% Starter	% Advanced
Fish feed production	0.019	1	0
Fish seed production	0.000	4.5	1.25
Market extension	0.000	11.5	35.25
Fish species extension	0.000	5.75	21.5
Product range extension	0.262	8.25	8.75
Utilization of other fish part	0.014	0	2

4 CONCLUSION

Based on the analysis of the data collected, it is suggested that small-scale fish farmers from groups with different levels of POKDAKAN showed significant differences in implementing business innovations. Different levels of fish farmer groups seem to have different interest in business innovations preferences. Fish farmers from advanced levels are more interested in implementing market extension and fish species extension, while starter groups prefer to produce their fish seed.

Three business innovations were found to be significantly different when carried out by small-scale fish farmers who belonged to different levels of fish farmer groups. These business innovations were fish seed production, market extension, and fish species extension.

REFERENCES

Antaranews. (2007) *Presiden Yudhoyono Memulai Kunjungan ke Jateng (President Yudhoyono started his Central Java visit)*, *Antaranews*. http://www.antara.co.id/print/?i=1171945075 (Accessed: 5 February 2019).

Carland, J. W., et al. (2007) "Differentiating entrepreneurs from small business owners: A conceptualization," in *Entrepreneurship*. Springer, pp. 73–81.

Central Java Fisheries and Marine Affairs Office. (2005) *Profil Potensi Perikanan Budidaya Jawa Tengah 2005*, Central Java Fisheries and Marine Office in Cooperation with the Fisheries and Marine Development and Research Institute. Central Java-Indonesia.

Central Java Statistical Center (2016) *Data penduduk Jawa Tengah 2014*, Central Java Statistical Center.

Jiansan, J., et al. (2001) "Aquaculture development beyond 2000: Global prospects keynote address II'. Bangkok (Thailand) FAO/NACA.

Kadin Sumbar. (2008) *Sekilas Tirta Mina Saiyo di Rimbo Binuang: kelompok petani ikan meningkatkan perekonomian*, http://www.kadin-sumbar.or.id/mod.php?=publisher&printarticle&artid=600 (Accessed: 5 February 2019).

Ministry of Marine Affairs and Fisheries. (2006) *The Department of Marine and Fisheries' Support in Overcoming Poverty*, The Department of Marine and Fisheries. Indonesia.

Pallant, J. (2016) "SPSS survival manual: A step by step guide to data analysis using SPSS 6th Edition."

Semarang Marine Affairs and Fisheries Office. (2005) *Pedoman pembentukan kelompok (Fish Farmer group formation guidebook)*, *City of Semarang Marine Affairs and Fisheries Office- Semarang*.

Suara Merdeka. (2002) *Petani ikan emas dan koi dibantu: ratusan ton terserang herpes*, *Suara Merdeka*. http://www.suaramerdeka.com/harian/0212/27/eko1.html (Accessed: 5 March 2019).

Suara Merdeka. (2007) *Budidaya ikan terkendala bibit*, *Suara Merdeka*. www.suaramerdeka.com/harian/0712/22/pan02.html (Accessed: 5 February 2019).

Facing Global Digital Revolution – Nirmala Arum Janie, Dwi Mulyaningsih & Wahyu Rachmawati (eds)
© 2020 Taylor & Francis Group, London, ISBN 978-0-367-33912-8

The value increase of Indonesian manufacturing companies within the period of 2014–2016, mediated by the capital structure

Emmy Susiatin, Kesi Widjajanti & Endang Rusdianti
Universitas Semarang, Semarang, Indonesia

ABSTRACT: Company value is essential for companies going public as reflecting the companies' performance, which may influence the perceptions of prospective investors and their decision to invest in those companies. High stock price may result in prospective investors perceiving better company conditions. Based on previous research, some factors influencing the company value include profitability, liquidity, capital structure, and growth opportunity. This research examines company value increases with structural capital as the intervening variable of the manufacturing companies listed on Indonesian Stock Exchange (IDX) during the period of 2014–2016. Secondary data, obtained from financial statements of the manufacturing companies, are analyzed using SEM and SmartPLS 2.0. The results reveal that profitability affects capital structure, with a significance value of 0.198; liquidity does not affect capital structure with a significance value of 13.132; growth opportunity affects capital structure with a significance value of 0.170; profitability does not affect company value with a significance value of 4.470; growth opportunity does not affect company value with a significance value of 1.996; liquidity affects company value with a significance value of 0.452; capital structure does not affect company value, with a significance value of 2.138; yet liquidity indirectly affects company value with capital structure as the intervening variable.

Keywords: profitability, liquidity, capital structure, growth opportunity and company value

1 INTRODUCTION

Research on factors affecting company value have been previously conducted, yet the results are still inconsistent. From the existing empirical facts, some previous research shows that there are some research gaps. For example, the research conducted by Hermuningsih (2013) shows that capital structure positively affects company value, while Welley and Untu (2015) argue that capital structure doesn't affect company value. In addition, the research conducted by Hermuningsih (2013) shows that profitability positively affects company value, while Mitta (2017) argues that profitability does not affect company value. Furthermore, the research conducted by Anzlina (2013) shows that liquidity positively affects company value, while Glos, Steade, and Lowry (1976) state that liquidity does not affect company value. The research conducted by Hermuningsih (2013) shows that growth opportunity positively affects company value, while Lilik S (2015) states that growth opportunity does not affect company value. The research conducted by Hermuningsih (2013) also shows that profitability negatively affects capital structure, while Glos, Steade, and Lowry (1976) state that profitability does not affect capital structure. The research conducted by Paramitha (2011) shows that liquidity positively affects capital structure, while Rizky Dian (2015) states that liquidity does not affect capital structure. Last but not least, the research conducted by Hermuningsih (2013) shows that growth opportunity positively affects capital structure, while Yoga (2016) states that growth opportunity negatively affects capital structure and Astuti (2015) finds that growth opportunity does not affect capital structure.

Research gaps found in the previous researches above – which use the variables of profitability, liquidity, growth opportunity, capital structure and company value – may then be solved using an intervening variable. In this research, capital structure is used as the intervening variable since the research is intended to see whether or not capital structure might mediate the effects of profitability, liquidity, and growth opportunity on company value. In addition, the research is conducted because the previous researches have shown that all variables affect each other, yet the role of capital structure in mediating the effects on company value is still less frequently examined by the researchers.

Manufacturing companies are considered as one sector of companies going public listed on the Indonesian Stock Exchange (IDX), consisting of the basic industrial and chemical sector, various industrial sectors, and the consumer goods industrial sector. Some manufacturing companies have experienced increases and decreases in company value. In the period of 2014 to 2016, 131 manufacturing companies listed on the IDX have experienced company value decrease. The company value decrease has been proxied by Price to Book Value (PBV) obtained from the PBV mean value of 126 companies that issued their financial statements in the period of 2014 to 2016.

2 LITERATURE REVIEW

Companies with high ROA (return on asset) tend to have a relatively small debt since high ROA provides a comparatively large internal fund. On the other hand, companies with lower ROA tend to use a larger debt because the available internal fund is relatively small and inadequate to fund the company's operations. Companies prefer using funds from internal capital; that is, funds coming from cash flow and retained earnings. The more profits that a company earns, the more available internal funds, meaning that the company does not need to have any debt. Similarly, according to Hermuningsih (2013), companies with high ROI (Return on Investment) tend to have a relatively smaller debt since high ROI enables companies to use the internal fund obtained from the retained earnings. Companies with high profitability have more retained earnings or dividends than those with low profitability. On the other hand, companies with low ROI tend to first use their internal fund then cover their shortage be acquiring debt. With high profitability, companies may have greater internal funds, and the composition of internal capital may also become greater than the use of debt.

Based on the theories above, the greater the ROA, the smaller the debt/equity ratio (DER) showing that profitability negatively affects capital structure. This theory is consistent with the research conducted by Hermuningsih (2013) and Astuti (2015) stating that profitability negatively and significantly affects capital structure. Thus, the research hypotheses are formulated as follows:

H1: Return on assets affects debt to equity ratio

Liquidity measured with current ratio (CR), is the ability of a company to fulfill all short-term liabilities due using its short-period possessed assets. Greater CR shows that the company has greater ability to fulfill its short-term liabilities. The research conducted by Glos, Steade, and Lowry (1976) found that company liquidity negatively affects capital structure. This condition reflects that higher CR may lower the company's capital structure (DER). It means that the company is able to pay its debts (external funding). Companies with high CR are more interested in first using their internal funds before turning to the external funding through debt; high CR means that those companies have excessive current assets which are sufficient to finance the company's operations without borrowing from external parties. In addition, the higher the company's CR level, the lower the company's debt that eventually affects the DER to decrease. Thus, the second hypothesis of this research is as follows:

H2: Current ratio affects debt to equity ratio

Growth opportunity, measured with price earnings ratio (PER), is the change of total assets belonging to the company (Kartini, 2008, in Yoga, 2016). Companies with high PER are supposed to increase their fixed assets to keep up with the company's growth. It means that the companies will need more funding and more retained earnings in the future. Mai (2006) in Hermuningsih (2013) states that retained earnings or dividends of companies with high PER may increase, and that the companies tend to maintain the target debt ratio. Thus, the third hypothesis of this research is as follows:

H3: Price earnings ratio affects debt to equity ratio

High ROA levels show that the company has better prospects so investors are interested in buying the company's stocks and company management may easily draw more capital to extend its businesses (Hermuningsih, 2013). If the company's stock demand increases, the company's stock price in the capital market may also indirectly increase. Hermuningsih (2013) and Glos, Steade, and Lowry (1976) show that profitability positively and significantly affects company value. Thus, the fourth hypothesis of this research is as follows:

H4: Return on assets affects price to book value

Higher company liquidity shows a company's greater ability to fulfill short-term current liabilities. It means that the companies are considered able to pay their debts (external funding) and this increases investor trust and willingness to invest funds in the companies and provide opportunities for the companies to grow that eventually result in the increase of company value. One analytical instrument used to assess company liquidity conditions is current ratio. The research conducted by Anzlina (2013) shows that liquidity positively affects company value. Thus, the fifth hypothesis of this research is as follows:

H5: Current ratio affects price to book value

Companies with high growth opportunity (GO) are encouraged to use equity as their funding sources to avoid agency cost between stockholders and company management. Meanwhile, the companies with low GO are encouraged to use their debts as their funding resources since those oblige the companies to pay the interest regularly. Sartono (2001) in Hermuningsih (2013) explains that growth potential can be measured with research and development costs. The bigger the research and development costs, the better the company's prospect to grow. The research conducted by Hermuningsih (2013) and Pangulu and Maski (2014) show that GO positively and significantly affects company value. Thus, the sixth hypothesis of this research is as follows:

H6: Price earnings ratio affects price to book rvalue

Companies with large tangible assets tend to use bigger debt proportions than those with large intangible assets (Patonah, 2016). The research conducted by Hermuningsih (2013) shows that debt addition policy is a positive signal for investors and may affect company value. By having debt, the companies may control the excessive use of cash funds made by management. Fund cash control may increase managerial control that the company value may increase, as reflected in its stock price increase. Having strong confidence in the company's prospects in the future and stock price increase, company managers should use more debt as a positive signal trusted by the prospective investors. The research conducted by Hermuningsih (2013) and Pangulu and Maski (2014) shows that capital structure positively affects company value. Thus, the seventh hypothesis of this research is as follows:

H7: Debt to equity ratio affects price to book value

3 RESULT

This study uses 126 financial statements of 42 manufacturing companies listed on the Indonesian Stock Exchange between 2014 and 2016 and then analyzed the data using SEM and SmartPLS 2.0 program. The results show that profitability affects capital structure, with a significance value of 0.198; liquidity does not affect capital structure, with a significance value of 13.132; growth opportunity affects capital structure, with a significance value of 0.170; profitability does not affect company value, with a significance value of 4.470; growth opportunity does not affect company value, with a significance value of 1,996; liquidity affects company value with a significance value of 0.452; and capital structure does not affect company value, with a significance value of 2.138. However, liquidity indirectly affects company value with capital structure as the intervening variable.

REFERENCES

Anzlina, C. W. (2013) "Pengaruh tingkat likuiditas, solvabilitas, aktivitas, dan profitabilitas terhadap nilai perusahaan pada perusahaan real estate dan property di bei tahun 2006–2008'.

Astuti, R. P. (2015) "Pengaruh Profitabilitas, Size, Growth Opportunity, Likuiditas dan Struktur Aktiva Terhadap Struktur Modal Bank (Studi Pada Perusahaan Perbankan Di BEI Tahun 2009-2013)." *Journal Of Accounting*, 1(1).

Glos, R. E., Steade, R. D., and Lowry, J. R. (1976) *Business, Its Nature and Environment: An Introduction.*

Hermuningsih, S. (2013) "Pengaruh profitabilitas, growth opportunity, struktur modal terhadap nilai perusahaan pada perusahaan publik di Indonesia." *Bulletin of Monetary Economics and Banking.* Bank Indonesia, 16(2), pp. 1–22.

Mai, M. U. (2006) "Analisis Variabel-Variabel yang Mempengaruhi Struktur Modal Pada Perusahaan-Perusahaan LQ-45 di Bursa Efek Jakarta." *Ekonomika*, pp. 228–245.

Pangulu, A. L., and Maski, G. (2014) "Pengaruh Profitabilitas, Growth Opportunity, dan Struktur Modal Terhadap Nilai Perusahaan (Studi Pada Perusahaan Perbankan yang Terdaftar di BEI Periode 2011-2013)." *Jurnal Ilmiah Mahasiswa FEB*, 3(1).

Paramitha, F. (2011) *Analisis Faktor-faktor yang mempengaruhi Struktur Modal serta Pengaruhnya terhadap Harga Sahama pada Perusahaan LQ45 Periode Tahun 2006-2009.* Bandung: Seminar Nasional Teknologi Informasi & Komunikasi Terapan 2011 (Semantik 2011).

Welley, M., and Untu, V. (2015) "Faktor-Faktor Yang Mempengaruhi Nilai Perusahaan Di Sektor Pertanian Pada Bursa Efek Indonesia Tahun 2010–2013." *Jurnal EMBA: Jurnal Riset Ekonomi, Manajemen, Bisnis dan Akuntansi*, 3(1).

Facing Global Digital Revolution – Nirmala Arum Janie, Dwi Mulyaningsih & Wahyu Rachmawati (eds)
 ISBN 978-0-367-33912-8

The determinant of earnings management: Evidence from manufacturing companies listed on the Indonesia Stock Exchange (IDX)

Winarsih & Shunniya Mega Rahma
Sultan Agung Islamic University, Semarang, Indonesia

Dyah Nirmala A. Janie
Universitas Semarang, Kota Semarang, Indonesia

ABSTRACT: This study aims to determine the effect of information asymmetry, firm size, level of financial statement disclosure, and litigation risk on earnings management from manufacturing companies listed on the Indonesian Stock Exchange (IDX). The data used is secondary data, originally from the Indonesian Stock Exchange (IDX). The study included 47 manufacturing companies in one year and 141 manufacturing companies in three years, from 2014 to 2016. Determination of the sample used a purposive sampling method. Data analysis techniques used multiple linear regression analysis with IBM SPSS 25. The results of the study indicate that information asymmetry and firm size have no significant positive effect on earnings management, level of financial statement disclosure has a significant positive effect on earnings management, and risk litigation has a significant negative effect on earnings management.

Keywords: information asymmetry, firm size, level of financial statement disclosure, risk litigation, earnings management

1 INTRODUCTION

One of the indicators used to assess a company's financial performance is the profit generated by the company. The information contained in financial statements is expected to assist creditors and investors in making decisions related to the funds they invest. The tendency to see profit allows for behavioral deviations that management can do in terms of presenting financial statements.

Generally, companies use accounting principles with accrual based on financial statements with the aim to adjusting cash flow for better mirroring performance and company position (Radzi et al., 2011). Financial statements using accrual base can provide more complete and comprehensive information than those using cash base. However, the use of accrual accounting can provide an opportunity for management to freely choose the accounting method as long as it does not deviate from applicable financial accounting standards. Freedom to choose and replace this accounting method is what triggers engineering of financial information. The effort to influence this financial information is called profit management. In general, profit management is a manager's attempt to intervene or influence financial statements by altering information on financial statement components when recording and compiling information either for values that are not material, material, or even very material. Scott (2006) in Watiningsih (2011) defined profit management as action to select accounting policies by company management to achieve certain objectives.

Profit management is the act of management in the financial statement drafting process that can raise or lower the accounting profit in accordance with its interests (Scott, 1997). Meanwhile, according to Fisher and Resenzweig in Sulistyanto (2008), profit management is defined as actions undertaken by the manager to raise or lower the profit on the running period of the company without causing an increase or decrease in economy profit of the company in the long term. Meanwhile, other parties tend to consider engineering done by management is not cheating, as long as it is done using accepted accounting methods and procedures. Therefore, this research aims to discover the motivation that encourages managers to perform managerial engineering activities by doing profit management.

Based on the explanation above, this study conducted tests on factors affecting profit management, based on information asymmetry, company size, financial report disclosure rate, and litigation risk on manufacturing companies in Indonesia.

2 RESEARCH METHOD

The population of this research was all manufacturing companies listed on the Indonesia Stock Exchange and samples in this study used purposive sampling methods, with criteria: A manufacturing company listed on the Indonesia Stock Exchange (IDX) continuously during the observation period, produced annual reports consistently, published financial statements with Rupiah currency unit, and has not suffered a loss during the years 2014–2016. Data analysis techniques using multiple linear regression analysis with IBM SPSS 25.

3 RESEARCH RESULTS AND DISCUSSION

The data used in this study is 47 companies during three research periods from 2014 to 2016. It gained 141 studies sample used as the observation data.

3.1 *Descriptive statistics*

Descriptive statistics were used to describe or provide descriptions of data in a study that can be seen from the average value (mean) and standard deviation

Table 1. Descriptive statistics research variable description.

	N	Minimum	Maximum	Mean	Std. Deviation
DA	141	-0.28802	0.80383	-0.8188167	0.11428134
Information asymmetry	141	3.92157	174.24758	54.4934724	34.44558167
Company size	141	21.68488	33.19881	28.4681392	1.78535846
Disclosure level	141	30.85000	82.98000	57.6428369	10.20559751
Litigation risk	141	25.82595	299.95862	32.5338472	22.83413486
Valid N (listwise)	141				

Source: Processed secondary data, 2019

Based on Table 1 above, the five variables indicate the average value of the result is greater than the standard deviation, so that indicates a fairly good outcome. This is because standard deviation is a reflection of a very high deviation, so that the data dissemination shows normal results.

3.2 *Regression analysis*

Table 2. Regression equality model.

	Unstandardized coefficients		Standardized coefficients		
Model	B	Std. Error	Beta	T	Sig
Constant	-0.0470	0.0788		-0.5958	0.5526
Asymmetry information	8.150E-06	0.0001	0.0055	0.0573	0.9544
Company size	-0.0011	0.0028	-0.0422	-0.4111	0.6819
Disclosure level	0.0010	0.0005	0.2568	2.5820	0.0110
Litigation risk	-0.0004	0.0002	-0.2002	-2.0100	0.0470

Source: Processed secondary data, 2019

Based on Table 2 above, the regression equality is as follows:

$$Y = -0,0470 + 8,150E - 06X_1 - 0,0011X_2 + 0,0010X_3 - 0,0004X_4$$

4 HYPOTHESIS TESTING

4.1 *The effect of information asymmetry toward profit management*

Based on the results of the research, it can be noted that information asymmetry has no significant effect on profit management because of the possibility of an error in the previous financial statement that is inappropriate with qualitative rules (Sulistyanto, 2008, in Firdaus, 2013).

The result of this research is not in line with agency theory because agency theory explained that agent and principal have different interests. The results showed that information asymmetry has no significant effect on profit management due to errors in previous financial reporting inappropriate with qualitative rules. This means information owned by the manager is also potentially inaccurate, so the action opportunities, such as profit management that the manager will do, also cannot be achieved.

The research is in line with research conducted by Maiyusti (2014) and Barus, Setiawati, and Andreani (2015), which stated that information asymmetry has no significant effect on profit management, and differs from the research conducted by Manggau (2016) and Yustiningarti and Asyik (2017), which stated that information asymmetry significantly positively affected profit management.

4.2 *The effect of company level toward profit management*

Based on the results of this research, it can be noted that company size has no significant effect on profit management as company size does not become the sole consideration for investors in investment decision making: other factors in investment decision making include profit level, company business prospects in the future, and so on. Indonesian investors are speculative in nature and tend toward capital gain. Moreover, in Indonesia, companies with a magnitude of assets have not guaranteed good company performance (Faozi, 2003 in Lusi, 2014, in Husna, 2015).

This research is in line with research conducted by Manggau (2016) and Gunawan, et al. (2015), which stated that the size of the company has no significant effect on profit management, as compared with research conducted by Kumala (2016) and Tarigan (2011), which suggested that the size of the company affects profit management.

4.3 *The effect of financial statement disclosure level toward profit management*

Based on the results of the research, it can be noted that the level of statement disclosure has no significant effect on profit management when the company presents full and complete company financial information likely to cause excessive information. Excessive information is dangerous because the detailed presentation and unimportant information can actually obscure the significant information and make financial statements difficult to interpret.

The result of this research is not in line with agency theory, because high levels of disclosure will be harmful, because the detailed presentation and unimportant information obscure the significant information and make financial statements difficult to interpret, so that the imbalance of information owned between investors and managers will also be greater and the profit management practice is also increasingly greater.

This research is in line with research conducted by Puspita and Kusumaningtyas (2017), which stated that the level of financial statements disclosure has a positive effect on profit management, different with the research conducted by Rahmi (2017), which suggested that the level of financial statements disclosure negatively affects profit management.

4.4 *The influence of risk litigation toward earnings management*

Based on the results of this research it can be noted that the litigation risk has significant negative effect on profit management as litigation risk is also expected to suppress profit management practices. The company will try its best to minimize the existence of lawsuits, because when it happens it will lower the company's image in the eyes of people, investors, and creditors as well as other external parties.

The result of this research is in line with agency theory, where there are different interests among managers and shareholders, but litigation risk faced by the company will affect management behavior, because managers will attempt to avoid any risks.

This research is in line with research conducted by Sari (2015), which stated that litigation risk negatively affects profit management, a finding that differs from research conducted by Kirana, Hasan, and Hardi (2016) and Puspita and Kusumaningtyas (2017), which stated that the litigation risk has a positive effect on profit management.

5 CONCLUSIONS

Information asymmetry and company size no significant effect on profit management. Meanwhile, the level of financial statement disclosure has a significant positive effect on profit management and the litigation risk significantly negatively affects profit management. This research only uses manufacturing companies as samples of the research. The results of this research show information asymmetry variable, company size, the level of financial statements disclosure, and the litigation risk only affects for 5.7% and a period of only three years. Further studies can use other models in calculating profit management, such as healy models, de Angelo model, Jones models, etc.

REFERENCES

Barus, A. C., and Setiawati, K. (2015) "The influence of asymmetry information, corporate governance mechanism, and deferred tax expense on earning management." *Jurnal Wira Ekonomi Mikroskil*, 5.

Firdaus, I. (2013) "The influence of asymmetry information and capital adequacy ratio towards earning management." *Final Project: Universitas Negeri Padang.*

Gunawan, K. et. al. (2015) "The influence of company size, profitability, and leverage towards earning management of manufacturing companies listed in Indonesia Stock Exchange (IDX)." *E-Jurnal S1 Ak Universitas Pendidikan Ganesha*, 3(1).

Husna, N. Y. (2015) "the influence of profitability, leverage, age, and company size toward earning management." *Final Project: Universitas Islam Negeri Syarif Hidayatullah Jakarta.*
Kirana, R. A. H., and Hardi. (2016) "The influence of financial statement disclosure, managerial skills and litigation risk toward earning management with audit quality as moderator variable." *Jurnal Akuntansi*, 4(2).
Kumala, I. (2016) "The influence of corporate governance, company size, and leverage toward earning management (A study at Food and Beverage Company Year 2012–2014)." *Sidoarjo: Universitas Muhammadiyah Sidoarjo.*
Kurniawati, N. (2011) "The influence of financial statement disclosure toward earning management wuth audit quality as moderator variable." *Universitas Brawijaya Malang.*
Maiyusti, A. (2014) "The influence of asymmetry information, managerial ownership and employee stock ownership program toward earning management." *Final Project: Universitas Negeri Padang.*
Manggau, A. W. (2016) "The influence of asymmetry information and company size toward earning management on mining companies listed in Indonesia Stock Exchange." *Jurnal Ekonomi dan Keuangan*, 13(2).
Ningsih, R. S. (2017) "The influence of disclosure toward earning management." *JOM Fekon*, 4(1).
Puspita, E. and Kusumaningtyas, D. (2017) "The Influence of ownership managerial mechanism, managerial skill, financial statement disclosure level toward earning management with audit quality as intervening variable." *Efektor: Universitas Nusantara PGRI Kediri.*
Sari, A. P. (2015) "The influence of litigation risk towards toward earning management with audit quality as moderating variable." *Fakutas Ekonomi Universitas Negeri Padang.*
Sulistyanto, S. (2008) "Earning management, theory and model empiris." *Jakaerta: Grasindo.*
Tarigan, T. C. (2011) "The influence of asymmetry information, corporate governance, and company size towards earning management practice." *Yogyakarta: Universitas.*
Yuskar, N. D. (2014) "The analysis of intellectual capital infulence towards company value with financial performance as intervening variable in banking companies of Indonesia." *Jurnal Manajemen dan Bisnis Sriwijaya*, 12(4).

Facing Global Digital Revolution – Nirmala Arum Janie, Dwi Mulyaningsih & Wahyu Rachmawati (eds)

Identification of the potential market segment of traditional medicine products based on the level of preference, satisfaction and risk perception

Rizka Zulfikar, Prihatini Ade Mayvita, Purboyo & Hj. Farida Yulianti
Universitas Islam Kalimantan MAB, Banjarmasin, Indonesia

ABSTRACT: This study aims to describe the potential market segments of traditional medicine products demographically by measuring the level of preference, satisfaction, and perceptions of community risk. This can be used by traditional medicine product manufacturers to determine the target market. The research was conducted in Banjarmasin starting from October 2018 to March 2019 and the population consisted of the Banjarmasin community, with up to 150 respondents. The sampling method used was purposive sampling, while data was collected using a questionnaire and data editing was carried out. Furthermore, this study uses qualitative - quantitative descriptive analysis and average comparison test. The results of the study found that: (1) Preference level, satisfaction and risk perceptions of the Banjarmasin community in terms of traditional medicine products were fairly good because the respondents' evaluations were in the middle category; (2) the potential demographic segment is used as a target market for green products, namely, the female gender, ages between 21-30 years, occupation comprising Government employees and Students, income between Rp. 1 - Rp. 4.9 Million, and education level from elementary to undergraduate.

Keywords: preference, satisfaction, risk perception, traditional medicine

1 INTRODUCTION

Some of the green products currently being developed are the products in the traditional medicine sector (TM) which consists of herbal products, standardized traditional medicines, and phytopharmaca. This is one of the mainstay sectors driving the Indonesian economy. Data from the Indonesia BPOM (National Agency of Drug and Food Control) shows that there are at least 751 businesses that produce traditional medicines and 626 of them are middle, small and micro enterprises (MSMEs) (BPOM-RI, 2018). Based on data from health research in 2017, almost half (49.53%) of the Indonesian population are in the age group of 15 years and above, while the proportion that consumes herbal medicine daily is 4.36%, and the rest (45.17%) consume it occasionally. The types of herbs mostly chosen for consumption include liquid medicine with a proportion of 55.16%; powder (43.99%); and brewed herbs (20.43%). Meanwhile, the smallest proportion is a modern packaged herbal medicine in the form of capsules/pills/tablets (11.58%) (BPOM-RI, 2018).

Through segmentation, companies divide large and heterogeneous markets into smaller segments which are effectively reached with unique products and services according to market needs (Kotler and Armstrong, 2012). The use of this strategy has been widely developed in business activities and is used to better serve customers, analyze consumer behavior and design products (Kasali, 2007).

In order to further develop products in the traditional medicine sector, this study was conducted to determine the potential market segments of green products by measuring the level of preference, satisfaction and public risk perceptions which can then be used by the product developers for marketing policies. Based on this background, the problem statement was developed as follows:

- What are preferences, satisfaction and risk perceptions levels of the Banjarmasin community on traditional medicine products demographically?
- What are the potential segments of traditional medicine consumers?

2 LITERATURE REVIEW

Marketing segmentation is an attempt to divide the market into groups that can be distinguished from each other in terms of needs, characteristics, or behavior which might require certain products and marketing strategies to reach them (Kotler and Armstrong, 2012). This method is also used as a means of understanding the market structure.

Factors such as gender, age, income level and type of work influence the satisfaction of green products in order to increase their buying interest (Schiffman and Kanuk, 2008; Wadi and Rahanatha, 2013).

- Gender factors: Women tend to be more positive towards green products than men (Lee, 2009; Zulfikar and Mayvita, 2017) and the purchase behavior of these products is a more relevant concept for women because they tend to have a greater concern for environmental issues (Lee, 2009; Zulfikar and Mayvita, 2018).
- Age factor: The buying behavior of the age group 18 to 24 uses the information contained in a product to make purchases, while there is a significant indication that information on a product greatly influences the purchase behavior of those between 25 to 44 years (Dumanovsky et al, 2010; Zulfikar and Mayvita, 2017).
- Occupation: Consumers who have certain types of work generally consume different items compared to those with other types of work (Kasali, 2007).
- Income Level Factors: Income levels significantly influence the satisfaction of green products and are the most dominant variable in influencing purchasing decisions (Rezvani *et al.*, 2013; Wadi and Rahanatha, 2013; Girard, 2010).
- Education Level Factors: Tsakiridou, Mattas and Bazoche (2012) in his study found that the level of education had a significant effect on the purchase of green products for consumers in Malaysia because better understanding and knowledge come with a good level of education, and these become the basis for consideration in purchasing green products. This is in line with the opinion expressed by Kim and Seock (2009) that knowledge has an important role in attitudes of consumer acceptance because those with knowledge of environmental issues and product benefits tend to prefer these products even though the price is relatively expensive.

However, there are also some findings that are contrary to such prior research, for instance, gender differences do not mediate knowledge variables in influencing consumer satisfaction and even though there is a tendency for knowledge, consumers prefer green products to others (Utami, Gunarsih and Aryanti, 2014). Also, there is a positive but insignificant difference between male and female consumer satisfaction, although it is still dominated by female consumers (Yasa and Ekawati, 2015). Contributions of age, education, occupation, income, and family size variables are not significant to purchasing decisions and are likely influenced by other variables (Yahya, 2011).

3 RESEARCH METHODS

As a survey research, the population and sample comprises the Banjarmasin people and 150 respondents, respectively. This research uses data analysis techniques such as (1) descriptive qualitative-quantitative analysis, (2) descriptive analysis - variable categorization and (3) comparing Means test.

4 RESULTS AND DISCUSSION

4.1 *Characteristics of respondents*

The characteristics of the respondents used in the study are presented in Table 1 below:

Table 1. Characteristics of respondents.

	Respondents			Respondents Demographics Factor	
Demographics Factor	Freq	%	Demographics Factor Freq	Freq	%
Gender			Entrepreneur	32	21
Male	61	41	Professional	6	4
Female	89	59	Income		
Age			< Rp 1 Million	18	12
< 20 Years	26	17	Rp 1– Rp 2,49 Million	27	18
21 – 30 Years	29	19	Rp 2,5– Rp 4,9 Million	39	26
31 – 40 Years	34	23	Rp 5– Rp 9,9 Million	34	23
41 – 50 Years	36	24	> Rp 10 Milion	32	21
> 50 years	25	17	Education		
Occupation			Elementary/Junior/High School	36	24
Unemployed	7	5	Diploma (D1,D2, D3)	4	3
Student	26	17	Under Graduate	64	43
Gov. Employee	34	23	Graduate	36	24
Private Employee	45	30	Post Graduate	10	7

Source: Primary Data, Processed, 2019

4.2 *Descriptive analysis and variable categorization*

The results of the categorization of variables based on average values and standard deviations are presented in Table 2 as follows:

Furthermore, the results in Table 2 are explained as follows:

Table 2. Categorization variables.

			Respondent Assessment		
Variables	Mean	Standard Deviation	High	Middle	Low
Preferences	52.6	5.39	17%	69%	14%
Satisfaction	0.79	4.58	52%	43%	5%
Risk Perception	16.57	3.75	25%	44%	31%

1. 17% of the respondents gave an assessment of the preference variables in the high category, 69% in the middle category, and 14% were found in the low category. The data shows that the majority of respondents were in the middle category, therefore, it can be concluded that the level of the community's preference for traditional medicine products was fairly good.
2. 52% of the respondents gave an assessment of the satisfaction variable in the high category, 43% in the middle category, and 5% were in the low category. The data shows that most respondents assessed the satisfaction variable in the high category, thus, it can be concluded that the level of community satisfaction with traditional medicine products was very good.
3. There were 25% of respondents that gave an assessment of the risk perception variables in the high category, 44% respondents in the middle category, and up to 31% were in the low

category. The data shows that the majority of respondents assessed the risk perception of traditional medicine products as being in the middle category. This means that the people considered the risk of consuming traditional medicine remained fairly low.

4.3 *Compare means analysis*

The average of all demographic segmentation groups was compared to the variables of preference, satisfaction and risk perception of the community towards traditional medicine products. Thus, the results are presented in Table 3.

Based on the table, the variables are described as follows:

Table 3. Results of demographic comparison analysis.

Demographics	Preference	Satisfaction	Risk
Gender			
- Male	2.93	0.43	3.08
- Female	3.10	0.51	2.84
Age			
- Under 20 years	3.08	0.46	3.12
- 21 – 30 Years	3.14	0.55	2.69
- 31 – 40 Years	3.09	0.53	2.85
- 41 – 50 Years	2.94	0.44	3.06
- Above 50 years	2.92	0.36	3.00
Occupation			
- Unemployee	3.00	0.43	3.29
- Student	2.92	0.62	2.96
- Gov. Employee	3.94	0.38	2.71
- Private Employee	3.07	0.44	3.09
- Enterpreneur	3.16	3.16	2.81
- Professional	3.17	3.17	3.33
Income			
- Under Rp 1 Million	3.19	3.19	3.33
- Rp 1 – Rp 2,49 Million	3.39	3.39	2.78
- Rp 2,5 – Rp 4,9 Million	2.90	2.90	2.77
- Rp 5 – Rp 9,9 Million	2.94	2.94	2.97
- Above Rp 10 Million	2.97	2.97	3.03
Education			
- Elementary/Junior/High	2.81	2.81	2.86
- Diploma	3.25	3.25	3.00
- Under Graduate	3.43	3.43	2.89
- Graduate	2.97	2.97	3.08
- Post Graduate	3.04	3.04	3.00

Source: Primary Data, processed, 2019

1. In terms of gender, women tend to have a better level of preference, satisfaction and risk perception of traditional medicine products compared to men. This is in line with the findings of Lee (2009), Yasa and Ekawati (2015) and Zulfikar and Mayvita (2017) which state that the female gender tends to have a more positive assessment.
2. In term of age, the group of 21-30 years tends to have a higher level of preference, satisfaction and perception of traditional medicine products than the other age groups.
3. In regard to occupation, respondents who are government employees have a better level of preference and risk perception than other occupational groups, while students tend to have greater levels of satisfaction.
4. Concerning income, the group of respondents with an income level of Rp. 1 million - Rp. 2,499 million have a higher level of preference and satisfaction. While in terms of risk

perception, groups with an income level of Rp. 2.49 - Rp. 4.99 million have a greater level of risk perception for traditional medicine products. Thus, it can be posited that the potential consumers for these products are in the Rp. 1 million - Rp. 4.9 million income group.

5. Regarding education, respondents with an undergraduate education level tend to have higher preference and satisfaction levels than other groups while those with elementary/junior/senior high school education have the highest satisfaction.

Based on the highest results of respondent evaluations on demographics, the groups which can be used as market potential for traditional medicine products are the female gender, age group of 21-30 years, the occupations of government employee and students, income between Rp. 1 million - Rp. 4.9 Million, and education level from elementary to undergraduate.

5 CONCLUSION

From the results of the data analysis and research findings, some conclusions are made:

1) The preference, satisfaction, and risk perception levels of the Banjarmasin community on green products in the traditional medicine sector are fairly good because the majority give an assessment in the middle category.
2) Potential groups that can be used as the target market for traditional medicine products are the female gender, the age group of 21-30, government employees and students, income between Rp. 1 Million - Rp. 4.9 Million, and the elementary to undergraduate education level.

REFERENCES

Girard, T. (2010) 'The role of demographics on the susceptibility to social influence: A pretest study', *Journal of Marketing Development and Competitiveness*, 5(1), pp. 9–22.

Kim, S. and Seock, Y. (2009) 'Impacts of health and environmental consciousness on young female consumers' attitude towards and purchase of natural beauty products', *International Journal of Consumer Studies*. Wiley Online Library, 33(6), pp. 627–638.

Kotler, P. and Armstrong, G. (2012) 'Principles of Marketing. Edisi keempat belas', *Pearson-Prentice Hall: New Jersey*.

Lee, K.-H. (2009) 'Why and how to adopt green management into business organizations? The case study of Korean SMEs in manufacturing industry', *Management Decision*. Emerald Group Publishing Limited, 47(7), pp. 1101–1121.

Rezvani, S. *et al.* (2013) 'Consumers' perceptual differences in buying cosmetic products: Malaysian perspective', *World Applied Sciences Journal*, 26(6), pp. 808–816.

Schiffman, L. and Kanuk, L. L. (2008) 'Perilaku konsumen', *Jakarta: Indeks*.

Tsakiridou, E., Mattas, K. and Bazoche, P. (2012) 'Consumers' response on the labels of fresh fruits and related implications on pesticide use', *Food Economics*. Taylor & Francis, 9(1–2), pp. 129–134.

Utami, R. D., Gunarsih, T. and Aryanti, T. (2014) 'Pengaruh Pengetahuan, Kepedulian dan Sikap pada Lingkungan Terhadap Minat Pembelian Produk Hijau', *Media Trend*, 9(2).

Wadi, H. and Rahanatha, G. B. (2013) 'Hubungan variabel demografi dengan respon konsumen terhadap iklan produk kopi merek TOP Coffee di Kota Denpasar', *E-Jurnal Manajemen*, 2(9).

Yahya (2011) 'Pengaruh Variabel Demografi Konsumen Terhadap Keputusan Pembelian Produk', *Jurnal Ekonomi*, 8(1), pp. 23–40.

Yasa, B. M. A. S. and Ekawati, N. W. (2015) 'Peran Gender dalam Menjelaskan Pengaruh Sikap dan Norma Subyektif terhadap Niat Beli', *E-Jurnal Manajemen*, 4(7).

Zulfikar, R. and Mayvita, P. A. (2017) 'Tingkat Kepercayaan dan Minat Beli Masyarakat Banjarmasin Terhadap Produk Hijau Berdasarkan Segmentasi Demografis', in *Proceeding of National Conference on Asbis*, pp. 410–426.

Zulfikar, R. and Mayvita, P. A. (2018) 'The Relationship of Perceived Value, Perceived Risk, and Level of Trust Towards Green Products of Fast Moving Consumer Goods Purchase Intention', *JEMA: Jurnal Ilmiah Bidang Akuntansi dan Manajemen*, 15(2), pp. 85–97.

Facing Global Digital Revolution – Nirmala Arum Janie, Dwi Mulyaningsih & Wahyu Rachmawati (eds)
© 2020 Taylor & Francis Group, London, ISBN 978-0-367-33912-8

The urgency of granting value added tax incentives on sugarcane molasses to encourage renewable energy development

T.P Muswati, N.B Utami & Indriani
Universitas Indonesia, Indonesia

ABSTRACT: Indonesian government launched an alternative energy di-versification to overcome the scarcity of fossil energy. The bioethanol industry produced from sugarcane molasses is an industry expected to have a prospective role in the provision of bio-fuels. However, the growth of the molasses industry is decelerated by various factors, including the absence of Value Added Tax (VAT) incentive. This study aims to analyze the imposition of VAT on molasses from the principle of economic growth and tax collection. This research uses a case study with a qualitative analysis of grounded theory. The results of the research note that the imposition of VAT on molasses has not fulfilled the economic growth principle and has not been considered to obtain VAT incentive. The government should be able to consider granting VAT incentive on molasses to encourage the development of renewable energy.

Keywords: tax policy, tax incentives, value added tax, renewable energy

1 INTRODUCTION

The projection of energy demand, issued by the Agency for the Assessment and Application of Technology (BPPT) shows that the final energy demand will continue to increase and is dominated by demand for fuel, particularly for industrial and transportation sectors (BPPT, 2017). The volume of Indonesian oil imports from 2010 to 2016 still increase. In 2016 shows the volume of oil import totally 43.890,60 net weight of thousands of tons (Central Bureau of Statistics, 2017). If the government continues to rely on imports and become a net importer country, it will endanger national energy security and sustainability (BPPT, 2017).

Since 2005, the government has sought to find alternatives with the issuance of Presidential Instruction Number 10 of 2005 (Inpres No.10/2005) on Energy Saving. To support the action, through Presidential Regulation No. 5 of 2006 (Perpres No. 5/2006) on National Energy Policy, the government seeks to develop alternative energy to ensure the security of domestic energy supply and to support national development through energy diversification. The alternative energy should be renewable one, available abundantly in Indonesia, consisting of geothermal energy, hydropower, mini & micro-hydro energy, bioenergy, solar energy, wind power, and marine energy.

Furthermore, the government issued Presidential Instruction No. 1 of 2006 (Inpres No.1/2006) on the Provision and Utilization of Biofuel as Other fuels; it is followed up by the Regulation of the Ministry of Energy and Mineral Resources (Permen ESDM) Number 32 of 2008 on the Supply, Utilization, and Procedure of Biofuel Trade as Other Fuels. The Permen ESDM, having undergone several amendments (lastly into Permen ESDM No. 12/2015), regulates mandatory stages of biofuel mixture with fossils fuel. Biofuels (biodiesel, bioethanol, and pure plant fuel) emerge as an alternative to new renewable energy (NRE) that can replace fossils fuel.

One of the biofuels is biodiesel that is utilized as a substitute for fossils-based-diesel; it has developed quite significantly. Although the amount of production has not been by the installed capacity, the biodiesel industry has been able to meet domestic demand, while the rest is exported (BPPT, 2016). Meanwhile, biofuels of bioethanol type, utilized as a gasoline mixture (premium and pertamax), was only produced in the period 2006-2009. The problems related to the hampered development of bioethanol can be divided into three aspects, namely aspects of production, markets, and policies.

The first aspect is caused by the high price of raw materials used to produce bioethanol. Raw material expenditure by producers may reach 40-60% of the production cost of bioethanol (Rukmaya, 2014). The second aspect includes bioethanol price that is higher than imported fuel price; hence, the consumer tends to choose imported fuel, rather than mix it with bioethanol. The third aspect, there is no adequate policy structure that regulates bioethanol, including the imposition of strict penalties for those who do not implement the mandate of biofuel mixture with fossils fuel. The existing policy is also considered not yet synchronous with those policies in various government agencies.

Related to the high cost of obtaining raw materials, it is among others caused by the need for large raw materials to produce bioethanol. One of the most potential bioethanol raw materials is sugarcane drops (Bappenas, 2015). To produce one liter of bioethanol, it takes four kilos of molasses. Molasses is a Sugarcane Derivative Product (SDP) that is subject to VAT submission under Law No. 42 of 2009 on the Third Amendment to Law No. 8 of 1983 on VAT and Sales Tax on Luxury Goods (VAT Law).

The VAT for delivery of molasses adds the initial cost of production that must be spent to obtain raw materials. Consequently, the bioethanol industry will be hampered by the supply of raw materials. Although VAT on Molasses delivery is not affected by the base-price of bioethanol sales due to the tax credit system, the high cost of obtaining raw materials becomes one of the barriers for the bioethanol industry to grow.

To overcome the problems related to raw materials, government support is required through a tax-related policy, particularly the provision of VAT Facilities. Therefore, this study aims to analyze the urgency of giving VAT incentives on the delivery of molasses by considering the principle of economic growth and efficiency.

2 LITERATURE REVIEW

Fiscal policy is an economic policy used by the government to manage or direct the economy to better or desired conditions by altering tax revenues and expenditures (Syadullah and Nizar, 2013). In line with the definition, Mankiw, Quah and Wilson (2012) assert that through a set of fiscal policy that includes government spending and taxation, governments can regulate aggregate revenue rates to overcome the state's economic instability.

The tax policy is part of the fiscal policy, which includes the tax system as well as tax collection. Based on four important characters of tax according to the classical paradigm–i.e., it can be imposed, levied by law, does not gain immediate benefits, and is used to perform the functions of the state–a tax policy should be designed as well as possible to adapt to existing changes (Rosdiana & Irianto, 2013). In designing ideal tax policy, we should pay attention to taxation principles in tax collection.

Public Accountants in the United States, namely the Tax Division of the American Institute of Certified Public Accountants (AICPA, 2001) issue the Guiding Principles of Good Tax Policy: A Framework for Evaluating Tax Proposals. AICPA recommends ten tax principles to be considered, namely equity and fairness, certainty, the convenience of payment, an economy in collection, simplicity, neutrality, economic growth and efficiency, transparency and visibility, minimum tax gap, and appropriate government revenues.

The results of this study refer to the principle of economic growth and efficiency, emphasizing that a tax system should not reduce or impede the national economic goals such as economic growth, capital planning, and international competitiveness. This principle can be achieved by the tax system by the principles and objectives of the economy and the legal basis that regulates taxation.

Tax incentives can be interpreted as the convenience provided by the government in terms of taxation. In his book entitled The Principles of Economics, Mankiw argues that the existence of a tax incentive policy may influence people's decisions in their choice (Mankiw, 2001). According to Tanzi and Zee (2013), generally in a tax system, there are several tax incentives, such as Exemption (exclusion of certain income or tax object), Deduction (reduction of partial allowance or expense of taxation), Tax Credits (tax credit as deduction of tax liabilities), Tax Relief (reduction or decrease in tax rate), and Tax Deferals (deferment or suspension of tax payments). In VAT system, there are two types of incentive, namely exemption and zero rate.

3 METHODOLOGY

The study uses qualitative approach. The data that has been collected are then processed and analyzed to answer the research questions through data analysis techniques. The data collected are qualitative data; and the data collecting technique are literature study and field study. Field studies were conducted through in-depth interviews with policy makers and bioethanol industry actors. The collected data is then analyzed to explain the urgency of VAT incentives on sugarcane molasses delivery, taking into account the principle of economic growth and efficiency.

4 DISCUSSION

Under Article 4 paragraph 1 of VAT Law, the delivery and import of molasses shall be subject to 10% VAT. Delivery of molasses has not received VAT incentives, hence the expenditure to obtain molasses raw material will be higher. Even though the government provided VAT incentives, the incentive would still seem incapable, however, to encourage the use of molasses raw materials for producing bioethanol.

In accordance with Article 16B of VAT Law, there are two incentive forms, namely VAT exemption, and VAT not collected (Zero Rate). The tax incentives to encourage the success of sectors of high priority economic activities on a national scale, as well as to encourage development and enhance competitiveness.

Currently, there is no provision of VAT incentives on the delivery of bioethanol raw material, namely molasses. The government can assist producers to meet the needs of bioethanol raw materials by providing VAT incentives. Nevertheless, there are three considerations for providing the VAT incentive on the delivery of molasses:

1) VAT incentives can be given if molasses are determined as certain strategic taxable goods.
2) The right incentive type should be determined between VAT exemption or VAT not collected/zero rate
3) Taking into account the status of entrepreneurs in the molasses industrial chain, whether having status as a taxable person or non-taxable person.

The aforementioned three considerations are related to the selection of appropriate VAT incentives under Article 16B of VAT Law. However, if the facility is deemed not maximal to be applied for the delivery of molasses, other forms of facilities may be considered. This is related to the discourse of reducing VAT incentives, or replacing it by other schemes, as has been applied to the biodiesel industry, i.e., by diverting the funds of palm levies as a subsidy to overcome the difference in production price with the price of vegetable diesel retail in the market. Thus, vegetable diesel can stay competitive with diesel from fossil fuels. This scheme is considered more effective in solving the problem, in order to avoid unnecessary target in the provision of incentives. Unnecessary target means that the supposed industrial target of incentive is not enjoying the benefit from the incentive.

The ideal VAT incentive policy should also pay attention to the tax principle. The principle of economic growth and efficiency is commonly used to oppose the principle of neutrality in tax policy. This principle is achieved by a tax system that is consistent with the economic principles and objectives of the tax enforcement jurisdiction. Economic growth and efficiency can be hampered by tax rules that support a particular industry (AICPA, 2001).

The explained conditions are not in line with the principle of economic growth and efficiency. The imposition of VAT on sugarcane drops (molasses) is one of the inhibitors of the bioethanol industry to develop. Therefore, VAT incentives should be considered to encourage the development of bioethanol industry with molasses raw materials. Although it is undeniable that the mandatory mixing of biofuels for fossil fuels also needs to be enforced.

If the bioethanol industry is fully supported by the government, among others through the provision of VAT incentives, it can support the achievement of national energy policies and can broadly promote economic growth and development and provide new jobs for the community. The policy related to the support of the bioethanol industrial growth in particular does not exist; only general incentives are available, like investment-related VAT incentives.

5 SUMMARY

The reason why the government has not provided VAT incentives for molasses delivery is that molasses have not been determined as certain strategic taxable goods. Should the government provide VAT incentive for molasses, producers can still credit the tax inputs on the acquisition of goods and services to produce molasses in order not to burden the selling price. The VAT policy of molasses delivery is not in line with the principle of Economic Growth and Efficiency. Therefore, the imposition of VAT on the delivery of molasses as raw material becomes one of the inhibitors for bioethanol industry to develop. If the government has not issued any VAT incentive for molasses, it is recommended to provide alternative subsidy schemes to support the development of new and renewable energy in Indonesia.

REFERENCES

AICPA (2001) 'Guiding Principles of Good Tax Policy: A Framework for Evaluating Tax Proposals', *American Institute of Certified Public Accountants.*

BPPT (2016) 'Indonesia's 2016 energy outlook: Clean energy technology development initiative', *Jakarta, Indonesia: Center for Energy Resources Technology and Chemical Industry (PTSEIK).*

BPPT (2017) 'Indonesia 2014 energy outlook: Clean energy technology development initiative', *Jakarta, Indonesia: Center for Energy Resources Technology and Chemical Industry (PTSEIK).*

Center for Administrative Studies (PUSKA). (2011). Academic review report of value added tax policy on delivery of port services for sea transport in international shipping lanes. University of Indonesia: FISIP.

Mankiw, N. G. (2001) *Principles of Economics 2nd* Edition. USA: Harcourt College Publisher.

Mankiw, N. G., Quah, E. and Wilson, P. (2012) 'Pengantar Ekonomi Makro, Principles Of Economics An Asian Edition'. Biro Bahasa Alkemis (Penerjemah). Jakarta. Salemba Empat.

Republic of Indonesia. Law Number 42 Year 2009 on Value Added Tax on Goods and Services and Sales Tax on Luxury Goods, (VAT and PPnBM).

Republic of Indonesia. Attachment I of Presidential Regulation No. 22 of 2017 on the National General Energy Plan.

Republic of Indonesia. Regulation of the President of the Republic of Indonesia Number 5 Year 2006 on National Energy Policy.

Republic of Indonesia. Regulation of the Minister of Energy and Mineral Resources Number 32 Year 2008 on the Provision, Utilization and Procedure of Biofuel as Other Fuel.

Republic of Indonesia. Regulation of the Minister of Energy and Mineral Resources Number 12 of 2015 on Third Amendment to Regulation of the Minister of Energy and Mineral Resources Number 32 Year 2008 regarding the Supply, Utilization and Procedure of Biofuel as Other Fuel.

Republic of Indonesia. Instruction of the President of the Republic of Indonesia Number 1 Year 2006 on the Supply and Utilization of Biofuel as Other Fuel.

Rukmaya, L. (2014) 'Policy of value added tax facility on delivery of molasses to support the national energy policy', *University of Indonesia: FISIP.*

Statistics, C. B. of (2017) *The volume of oil and gas exports and imports (net weight: thousand tons), 1996-2016*, https://www.bps.go.id/.
Syadullah, M. and Nizar, M. A. (2013) 'Fiscal Policy: Theory and practice in Indonesia', *Jakarta*: *Observation & Research of Taxation.*
Tanzi, V. and Zee, H. (2013) 'Economic Issues No. 27: Tax Policy for Developing Countries', *International Monetary Fund.*

Facing Global Digital Revolution – Nirmala Arum Janie,
Dwi Mulyaningsih & Wahyu Rachmawati (eds)
 ISBN 978-0-367-33912-8

Understanding international tourists' preference pattern in visiting tourist destinations in Yogyakarta

Damiasih, Christantius Dwiatmadja, John J.O.I. Ihalauw & Lilie Suharti
Faculty of Economics and Business, Universitas Kristen Satya Wacana Salatiga, Salatiga, Indonesia

ABSTRACT: The purpose of this study is to understand and analyze the pattern of choice of international tourists in visiting tourist destinations in Yogyakarta. The pattern of choice is generally influenced by various factors that could be divided into two significant factors; internal and external. This study is based on empirical observation. These influential forms become the preference of international tourists in visiting tourist destinations. This research method uses a sample of international tourists visiting Yogyakarta in 2018. This study uses quantitative and descriptive data analysis. The results of this study indicate that the pattern of choice of international tourists in visiting tourist destinations in Yogyakarta is influenced by novelty seeking, particularly the uniqueness of a region, including such elements as panorama, culture, society, and hospitality. The more unique and interesting the reality is seen to be, the more the interest of international tourists to visit Yogyakarta will be increased.

Keywords: international tourist, tourist destination, Yogyakarta, tourist preference.

1 INTRODUCTION

Tourism is a very big industry. The tourism sector can drive changes in society in various forms such as lifestyle, people's purchasing power, buying goods and services, and also social and economic changes. Dębski also states that tourism is a significant element of the economy worldwide; for numerous regions or countries it serves as an important source of income, setting the stage for development (Dębski, 2014).

The phenomenon that occurs in the trend of tourism in the world is the development of various kinds of tourism products. Thus, tourism experiences commodification in order to influence the interest of tourist visits. Commodification of tourism is the center of attention in tourism studies, especially the commodification of culture and tradition. Commodification of tourism occurs because the process of searching for authenticity becomes a major interest in modern tourism (Alfath & Permana, 2016). This search for authenticity is an important part of the interest of tourist visits. The interest of tourist visits is generally formed due to the driving and pulling factors. These driving and pulling factors are actually the internal and external factors that can influence the interest of tourists to travel. Both of these factors are complementary and form the character of tourist visiting interests (Utama, 2017).

The search for authenticity in tourism leads to novelty seeking as a commodification process in the field of tourism in general. Commodification is interpreted as the result of human work in the form of goods or services that are intentionally produced to be exchanged through other mechanisms or market mechanisms. Commodification is the process of making something that is not a commodity into something to be or to be treated as a commodity that can be traded or exchanged for a particular purpose or profit (Mulyanto, 2018). Commodification has penetrated the fields of education, culture, religion, body, desire, and even death. In the world of tourism, commodification is also known as touristification, which is an effort to make the community a tourist destination as a tourism product (Sari, 2016). Commodification is actually a concept that is not only concerned with the problem of the production of commodities or goods in the narrow sense of the economy about the goods being traded, but rather more than that, namely about how the goods are distributed and consumed (Fairclough, 1995).

Novelty seeking as an element encouraging interest in tourist visits has various forms including nature, culture, social, religious values, social, climate, access, tourism products, and other facilities provided or offered to tourists. At present, travel is a part of the lifestyle of modern society that has the potential to influence the purchasing power of the people. As in Spillane's (1987) view, tourism is an activity of traveling with the aim of getting pleasure, seeking satisfaction, knowing something new, or improving health. These travel activities lead to the distribution of goods and services and increase human activities to carry out economic transactions in accordance with the ability and capacity of capital owned by the community. Thus, tourism travel experiences a trend for today's modern society.

The phenomenon of trips by tourists from various countries is interesting. Visiting tourist attractions in the process of finding, discovering, and enjoying the intended tourist destinations is the focus of attention in this study. Therefore, the progress or decline of the development of the tourism industry in a country or region is determined by how many tourists visit there, using the number of tourist visits in general as a basic reference in analyzing the development and progress of tourism in a country, for example, in the Southeast Asia region. The Association of Southeast Asian Nations is often referred to as a "tourist paradise" in Asia because it has abundant wealth in terms of tourism. The main destinations for tourism in ASEAN are Indonesia, Thailand, Malaysia, Singapore, and the Philippines (Sabon et al., 2018). Foreign tourists need a variety of supporting facilities and infrastructure and this has become one of the factors that affect the level of visits of foreign tourists (Sri, 2013).

Yogyakarta is one of the regions that becomes a consideration and reference for leading tourist destinations. The Yogyakarta region has not received much attention as an area with strategic potential to be developed into a foreign tourist attraction. Yogyakarta as the city of students, culture, heritage, tolerance, and also the city of culinary experiences should be able to bring in foreign tourists in greater numbers. The existence of the new slogan "Special Jogja" is expected to be able to bring in more foreign tourists but has not yet played a significant role. The purpose of this study is to explore and cultivate the international tourist's preference patterns in visiting a tourist destination. This study is based on empirical observation. The methods that this study uses are qualitative and quantitative analysis. Research gap in this study are novelty seeking, lifestyle, and purchasing power, which can be triggered for (especially foreign) tourist interest but has not been researched yet. Based on findings in this research, novelty seeking can be important element influencing tourist interest to visit and choose a tourist destination.

2 RESULT AND DISCUSSION

Pattern of choice of tourist preference becomes important in tourism studies. Research and study about pattern of choice of tourists can be base to look at development and dynamics of tourism in a country. Generally, research to look pattern of choice of tourist preference uses variables of income, tourist attraction, and lifestyle. This research will look pattern of choice of tourist preference by using the variables of novelty seeking, lifestyle, and purchasing power. Novelty seeking becomes a conceptual framework to deepen understanding about pattern of choice of tourist preference in decision-making and motivation to visit a tourist destination.

Novelty seeking influences the interest of foreign tourist to visit Yogyakarta. Yogyakarta offers a variety of tourist destinations in the form of increasingly diverse natural tourism, both long-known and newly introduced. Meanwhile, novelty-seeking behavior in tourism is often closely associated with exploratory behavior, as well as a variety of other terms, including curiosity drive, sensation seeking, and variety seeking, which describes the desire to seek arousal stimuli (Lee & Crompton, 1992).

Previously well-known natural attractions include the beauty of Parangtritis beach with its mythological stories, beaches along Gunung Kidul district, panoramic views of Mount Merapi, and nature tourism in the Kaliurang region. In its development, Yogyakarta also offers natural attractions or new natural panoramas such as the lava tour after the Merapi eruptions in 2006 and 2010, Breccia cliffs, Pindul caves, and Mangunan nature tourism in Bantul region. This natural panorama is an attraction and of interest to foreign tourists visiting Yogyakarta. The results of data processing show that the panorama of nature, religion, culture, and social tourism have a positive influence on foreign tourist visits. Apparently, foreign tourists are influenced by novelty seeking, lifestyle, and purchasing power when they are searching for tourist attractions, and these shape interest in tourist visits. The lifestyle of foreign tourists – which has a tendency to look for something new – has an impact on increasing the purchasing power of tourists.

Novelty seeking is the starting point for the commodification process from the lifestyle and purchasing power of tourists in visiting certain tourist destinations. Yogyakarta as a tourist destination for foreign tourists has an important role in carrying capacity in the process of commodification of tourist interest because Yogyakarta provides a new atmosphere and experience, and offers new tourist attractions. Novelty seeking is referred to as *neophilia* (Greek), which reflects the love of what is new. Neophilia is known to vary from culture to culture and from individual to individual. Variation in the degree of novelty seeking can be explained by optimal stimulation level theory, which posits that an individual prefers a particular level of stimulation and that this level varies from one person to another (Mak, 2014). Lee and Crompton (1992) proposed that there are six dimensions of novelty construct: change of routine, escape, thrill, adventure, surprise, and boredom alleviation.

In general, facilities and infrastructure in Yogyakarta that attempt to meet tourist needs are quite good and have a positive value. Transportation access is also easy and affordable. The variety of transportation options become supporting factors because they make it easy for foreign tourists to visit tourist destinations in Yogyakarta. The attraction aspect in various previous studies also shows that quality facilities, services, and infrastructure have an important influence in encouraging tourists to visit these tourist destinations (Wiradiputra & Brahmanto, 2016). Swastuti and Pudjiarti (2018) show that novelty seeking is based on tourist attraction, and that quality of services and destination image have a significant influence on tourist satisfaction and interest in revisiting. Ibrahim and colleagues' (2009) study on travelling patterns and tourism preferences shows that, generally, international tourists prefer travel efficiency, fast services, good facilities, and a friendly social atmosphere. This study concluded that international tourists visiting Yogyakarta prefer novelty seeking for new experiences and new attraction on the region. Findings in this research show that novelty seeking becomes variable to trigger the interest of international tourists to visit a tourist destination. Novelty seeking in the context of the development of tourism studies becomes commodification because this variable has elements to influence psychological, cultural, and social conditions of tourists to make rational choices about tourist destinations already known and understood.

3 CONCLUSION

Based on the results of the research on international tourist preference patterns, novelty seeking is a commodification of interest in foreign tourist visits to Yogyakarta. It can be concluded that foreign tourist visitors are interested in visiting Yogyakarta because the tourist destinations offered in Yogyakarta are quite diverse including historical, natural, cultural, and artistic aspects.

REFERENCES

Alfath, E. D., and Permana, Y. S. (2016) "Festival 1000 tumpeng: Komodifikasi tradisi, pariwisata, dan "territoriality"di Gunung Kelud." *Masyarakat, Kebudayaan dan Politik*, 29(4), pp. 169–180.

Dębski, M. (2014) "Tourism habits and preferences: Comparative analysis in selected European countries." *Journal of Intercultural Management*. De Gruyter Open, 6(4–1), pp. 39–54.
Fairclough, N. (1995) *Media Discourse*. E. Arnold.
Ibrahim, Z. et al. (2009) "Travelling pattern and preferences of the Arab tourists in Malaysian hotels." *International Journal of Business and Management*, 4(7), pp. 3–9.
Lee, T.-H., and Crompton, J. (1992) "Measuring novelty seeking in tourism." *Annals of tourism research*. Elsevier, 19(4), pp. 732–751.
Mak, A. H. N. (2014) "Novelty, tourism." *Encyclopedia of Tourism*. Springer, pp. 1–3.
Mulyanto, D. (2018) *Geneologi kapitalisme: antropologi dan ekonomi politik pranata eksploitasi kapitalistik*. Resistbook.
Sabon, V. L. et al. (2018) "Strategi Peningkatan Kinerja Sektor Pariwisata Indonesia Pada ASEAN Economic Community." *Esensi: Jurnal Bisnis dan Manajemen*, 8(2), pp. 163–176.
Sari, N. P. R. (2016) "Komodifikasi dalam Industri Perhotelan di Bali." *Analisis Pariwisata*, 16(1), pp. 23–28.
Spillane, J. J. (1987) *Pariwisata Indonesia: sejarah dan prospeknya*. Kanisius.
Sri, A. A. P. (2013) "Faktor-faktor yang memotivasi perempuan sebagai pengelola pondok wisata di Kelurahan Ubud Kecamatan Ubud Kabupaten Gianyar." *Analisis Pariwisata*, 13(1), pp. 1–10.
Swastuti, E., and Pudjiarti, E. S. (2018) "A study of novelty-seeking-based revisiting interest at Ketep Pass Tourism Object in Magelang Regency." *Indonesian Journal Of Business And Economics*, 1(2).
Utama, I. G. B. R. (2017) *Pemasaran Pariwisata*. Penerbit Andi.
Wiradiputra, F. A., and Brahmanto, E. (2016) "Analisis persepsi wisatawan mengenai penurunan kualitas daya tarik wisata terhadap minat berkunjung." *Jurnal Pariwisata*, 3(2), pp. 129–137.

Facing Global Digital Revolution – Nirmala Arum Janie, Dwi Mulyaningsih & Wahyu Rachmawati (eds)
© 2020 Taylor & Francis Group, London, ISBN 978-0-367-33912-8

Optimal portfolio with single index cut-off model in LQ 45 stocks on Indonesia Stock Exchange

Y. Karliena
Lecturer of LP3I Cilegon, Banten, Indonesia

I.R. Setyawan
Lecturer of Department of Management, FE UNTAR, Jakarta, Indonesia

ABSTRACT: This study aims to examine the relevance of the single index cut-off model for investors on the Indonesia Stock Exchange (IDX). The population consists of all the public firms listed on IDX during 2011–2013 and the sample comprises LQ 45 companies. The method of sample selection was conducted through purposive sampling and there are two criteria: 1) the company is included in the top 10 highest liquidity of the LQ45 category, and 2) the company has never been one of the top 10 category. Using the single index cut-off model from Elton and Gruber (1997), drastic changes in the selected stock of investors from 2011 to 2013 were found. These indicate investor demographic changes that are caused by differences in the emerging trendsetters in the capital market, in line with the theory of fad and fashion in capital markets (Shiller, Fischer, & Friedman, 1984). This study also finds that ASII, UNVR, and ISAT are always included in the optimal portfolio, which shows the effectiveness of the single index cut-off model in previous studies.

Keywords: single index cut-off model, LQ 45, investors, IDX, risk avoider, risk seeker

1 INTRODUCTION

Sartono and Zulaihati (1998) researched selecting the optimal portfolio with the single index method, with emphasis placed on investor's rationality, in other words, the investor's ability to choose the optimum portfolio based on preference, risk, and behavior in setting up the transaction. The use of twenty-five LQ45 capital companies for the period 1994–1996 showed three portfolio candidates' capital, namely, Lippo Land Development (LPLD), Astra International (ASII) and Gudang Garam (GGRM) with compositions of 51.32%, 47.55%, and 1.13% respectively. Investor's rationality was proven using an independent t-test. Yuniarti (2010), in a research test based on Sartono and Zulaihati (1998), used bank's capital with adequate consideration given to the years prior to 1997, which mainly used only manufacture's capital. Meanwhile, for the bank capital of period 2009–2010, three flagship stocks were found: BBRI, BBCA, and BBNI with proportions 58.15%, 23.72%, and 18.13% respectively. As a continuation of Sartono and Zulaihati (1998) and Yuniarti (2010), Sembiring (2012) used the Panin Securities Investment Manager approach, specifically, the Panin Dana Maksima's Mutual Fund (PDM) for the period of March 2011. With the single index method, the author discovered five superb stocks: Duta Pertiwi (13.20%), Delta Djakarta (25.59%), Bank Niaga (52.01%), Panin Insurance (5.72%), and Kalbe Farma (3.49%). Furthermore, Ramanathan and Jahnavi (2014) tested the model for bank stocks and information technology in India's capital market for the period 2009–2013. There are five stocks included in the portfolio candidates: Dish (7.24%), Saregama (13.45%), Suntv (13.95%), HT Media (24.79%), and PVR (40.58%). Nalini (2014) attempts to prove the accuracy of the single index model with stocks in BSE for period 2009–2014. The findings show four stocks of candidate portfolio optimums: ITC (70.88%), Tata (10.08%), Dr. Reddy's L (17.41%), and Bajaj Auto (1.63%).

The single index cut-off model has been used by several researchers. However, in this study, the model is still considered relevant because it was used based on a few important basic elements, for instance, stock individual return. Interestingly, researchers in Indonesia focus on the industrial sector while in India they are more focused on the IT (Information Technology) sector. The difference is due to the investors' diversification strategies, meaning that India's strategy displays more homogeny. In Indonesia, heterogeneity in the industry sector still exists. Thus, the objectives of this research are to retest the single index cut-off model, which still existed in IDX from 2011–2013, in accordance with Elton and Gruber (1997), and selecting the stocks that are always present in the optimal portfolio.

2 RESULT OF RESEARCH

2.1 *Profile of average risk & return of 10 selected LQ45 stocks for 2011–2013*

Table 1 presents both return and risk for 10 preferred LQ45 stocks between 2011 and 2013. In 2011, the highest return was obtained by BBCA stocks and the lowest return by KLBF, while the highest risk goes to INCO and the lowest to UNVR stocks. Thus for 2011, the preferred stock for the risk-avoidant investor (type 1) is BBCA and for risk-seeker investor (type 2) is UNVR. While the preferred stock type for the risk-seeking investor (type 2) is KLBF and for the risk-avoidant investor (type 1) is INCO. The risk-avoidant investor (type 1) only wants the highest return, similar to the risk-seeking investor (type 2) who seeks the highest risk.

In terms of 2012 data, higher risk was obtained by KLBF stocks and higher returns by AALI. However, KLBF stocks obtain second place for the highest return. These findings confirm the theory of portfolio formation for investors who are risk-seekers. Meanwhile, if this theory is excluded, risk-seekers (type 2) who are confirmed to AALI stocks will also consider KLBF as an alternative in the portfolio pair. From the data in 2013, it appears that PTBA has the highest return and risk. Therefore, PTBA will be the first priority as a portfolio for risk-seeking investors. As a compartment in the stock market, risk-avoidant investors will select GGRM and as compensation for premium market risk then TLKM or BCA can be used as alternatives. These differences between risk seekers and avoiders are in accordance with Shiller, Fischer, and Friedman (1984) who posits that there are fads and fashion in IDX.

2.2 *Stock ranking profile based on ERB value of LQ45*

Selecting stocks using risk and return criteria has significant subjectivity because it is based only on the type of investor. Therefore, in Table 2, the process of selecting stocks is outlined based on ERB rank (excess return to beta), which can be compared to the Sharpe ratio.

Table 1. Profile of average risk and return 10 LQ45 selected stocks for 2011–2013.

Stock	Return 2011	Risk 2011	Return 2012	Risk 2012	Return 2013	Risk 2013
TLKM	0.118	2.437	0.068	1.957	0.163	2.159
AALI	0.133	2.470	0.178	2.152	0.289	2.587
BBCA	0.275	2.207	0.129	1.730	0.169	2.181
ASII	0.242	2.368	0.009	2.216	-0.072	2.887
GGRM	0.001	1.758	-0.090	2.302	-0.266	1.890
UNVR	-0.054	1.752	0.083	2.199	-0.234	2.046
KLBF	-0.014	3.296	0.168	2.544	0.177	2.775
PTBA	0.177	3.266	0.047	2.103	0.493	3.700
ISAT	0.240	2.390	-0.010	2.036	-0.140	2.379
INCO	0.045	3.859	0.032	2.170	0.276	2.985

Table 2. Stock rating based on ERB value for 2011–2013.

Stock	ERB 2011	C* 2011	ERB 2012	C* 2012	ERB 2013	C* 2013
TLKM	-233.193	-146.227	-754.969	-35.913	-237.310	-51.403
AALI	-623.108	-47.654	-1670.591	-6.449	1616.165	3.231
BBCA	-355.263	-88.484	513.655	45.818	-242.021	-52.241
ASII	-355.263	-136.769	-634.789	-17.717	-140.417	-86.285
GGRM	-411.036	-99.092	-902.586	-11.238	-668.183	-4.946
UNVR	-490.293	-29.051	-2123.301	-9.260	971.151	2.596
KLBF	-1855.792	-61.674	-289.414	-49.056	-106.540	-56.100
PTBA	-147.642	-60.615	-564.509	-56.109	685.311	1.781
ISAT	-409.776	-106.290	-213.826	-64.481	-382.654	-21.526
INCO	-444.335	-18.274	3126.963	4.609	-360.501	-9.068

For 2012, all stocks are also selected in the optimal portfolio given that the ERB values are much larger than C* (cut-off indices). Interestingly, BBCA and INCO have positive ERB values. This is because the increase in stock prices per year from both stocks exceeds the increase in risk-free asset value (Rf), that is, the BI rate. Similar to the data for 2011 and 2012, all stocks in 2013 have a large ERB, meaning they will all be included in the optimal portfolio.

2.3 *Determination of the optimal portfolio proportion*

Based on Figure 1, all LQ45 stocks will be selected as an optimal portfolio. The biggest percentage, owned by UNVR, is more than 19% and the smallest percentage, owned by INCO, is 0.3%. Optimal portfolio selection can also be mapped in Figure 1 through a pie chart. The use of this is driven by the success of Ramanathan and Jahnavi (2014) and Nalini (2014), who conducted research in the Indian capital market.

The results from Figure 1 indicate that the risk-avoidant investor type dominates on IDX, while two mining sector stocks (PTBA and INCO) have high business risks that makes them less attractive to investors. Furthermore, based on Figure 2, all LQ45 stocks are also selected in the optimal portfolio for 2012. The stocks that have the highest percentage are PTBA with a proportion of 16.3% and the lowest is GGRM with 5.5%. The findings in 2012 are clearly different from those in 2011 as PTBA stocks that were initially not in demand by investors were later in great demand. GGRM was ranked second among the preferred stocks of investors in 2011.

In 2013, the most preferred stock of investors was ASII (automotive sector) with 26.7% and the lowest-ranked stock was PTBA (mining sector) with 1.5%. The findings in Figure 3 indicate a change in investor-return demographics. If the risk-seeking investor type dominated the IDX making mining stocks a favorite in 2012, then in 2013 it returned to the risk-avoidant investor type dominating the IDX making blue-chip stocks such as ASII and UNVR the preferred stock. The findings in Figures 1-3 confirm Shiller, Fischer, and Friedman's (1984) theory of fad and fashion in capital market. There are differences in investor bias, both emotional and cognitive, resulting in changes in investor demographic types that do not occur automatically. This is because investors will need a basis for decision-making called frames.

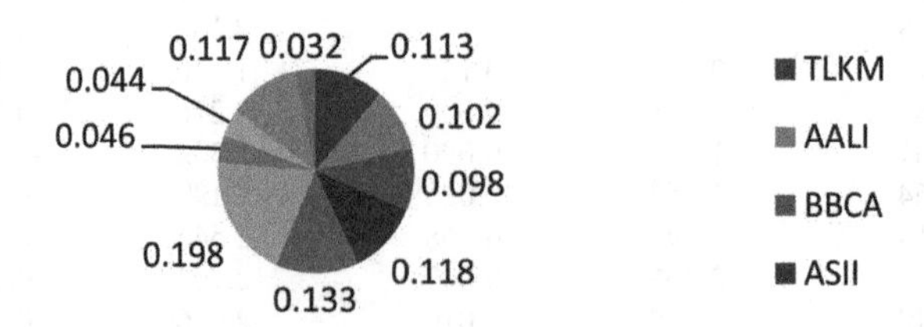

Figure 1. Graphic of optimal proportion portfolio in 2011.

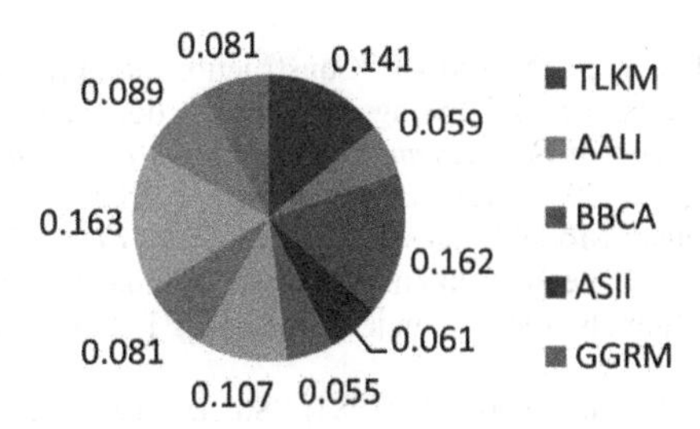

Figure 2. Graphic of optimal proportion portfolio for 2012.

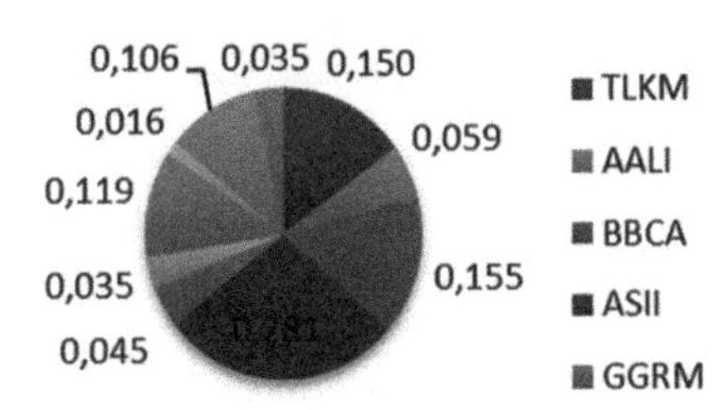

Figure 3. Graphic of optimal proportion portfolio for 2013.

This irrational type as the main basis of trendsetters is rooted in competition between concepts of EMH (efficient market hypothesis) and behavioral finance in terms of expected utility from investors for achieving investment objectives. However, based on the data from 2011–2013, it remains evident that the EMH theory is more dominant because, in principle, more investors are risk avoidant. This can be seen from the selection of blue-chip stocks in the highest order, such as ASII, UNVR stocks, and ISAT.

3 CONCLUDING REMARKS

The results of the analysis using 10 selected LQ45 stocks during 2011–2013 proved that the single index cut-off model can still be used properly. This model is able to provide an overview of the optimum proportion for all LQ45 stocks on the IDX without including the results of negative proportions (short selling). Evidence from the single index cut-off model shows rationality investors are still dominated by risk-avoidant types. In addition to the proven relevance of the model, this study also found that there was a change in the preferred stock of investors as seen in the drastic change in rank for a stock in the optimal portfolio proportion for each year. This indicates a demographic change in investors on the IDX, which occurred due to changing trendsetters in the capital market. According to Shiller, Fischer, and Friedman's (1984) fad and fashion in capital market theory, investors in the capital market are like consumers in the goods & services market. They will be subject to the spirit of favoritism for a particular product. In this case, the capital market will have an increasingly widespread choice of stock analyst versions.

REFERENCES

Elton, E. J. and Gruber, M. J. (1997), "Modern portfolio theory, 1950 to date." *Journal of Banking & Finance*. Elsevier, 21(11–12), pp. 1743–1759.

Nalini, R. (2014) "Optimal portfolio construction using Sharpe's single index model: A study of selected stocks from BSE." *International Journal of Advanced Research in Management and Social Sciences*. Citeseer, 3(12), pp. 72–93.

Ramanathan, K. V., and Jahnavi, K. N. (2014) "Construction of optimal equity portfolio using the sharpe index model with reference to banking and information technology sectors in india from 2009-2013." *International Journal of Business and Administration Research Review*, 2(3).

Sartono, R. A., and Zulaihati, S. (1998) "Rasionalitas Investor Terhadap Pemilihan Saham dan Penentuan Portofolio Optimal dengan Model Indeks Tunggal di BEJ." *Kelola*, 7(1998).

Sembiring, F. M. (2012) "Analisis pembentukan portofolio optimal berdasarkan Single Index Model pada saham-saham yang dikelola oleh manajer investasi PT Panin Securities." *Portofolio*, 9(1), pp. 1–17.

Shiller, R. J., Fischer, S., and Friedman, B. M. (1984) "Stock prices and social dynamics." *Brookings papers on economic activity*. JSTOR, 1984(2), pp. 457–510.

Yuniarti, S. (2010) "Pembentukan Portofolio Optimal Saham-Saham Perbankan dengan Menggunakan Model Indeks Tunggal." *Jurnal Keuangan dan Perbankan*. Universitas Merdeka Malang, 14(3), pp. 459–466.

Facing Global Digital Revolution – Nirmala Arum Janie, Dwi Mulyaningsih & Wahyu Rachmawati (eds)
© 2020 Taylor & Francis Group, London, ISBN 978-0-367-33912-8

A study on ASEAN–China capital market integration: An orthogonal GARCH model analysis

Christopher Kevin & Robiyanto
Universitas Kristen Satya Wacana, Salatiga, Indonesia

ABSTRACT: Capital market integration is the process of eliminating boundaries between countries in allocating financial assets such as stocks, bonds, and cash. This article assesses the degree of capital market integration among the ASEAN-5 countries and China as in recent years China has proven to be a force to be reckoned with in terms of economic development and has built relatively good trade relations with the ASEAN-5 countries. This article is also taking into consideration phenomena that might have a significant impact on the fluctuation of stock return that will affect the degree of integration among these countries. One of the phenomena that is taken to consideration is the United States' Presidential Election in 2016. This study resulted in the discovery of a significant degree of integration among Indonesia, Malaysia, Singapore, the Philippines, and China's capital market, whereas Thailand's capital market (SET) tends to be more segmented.

1 INTRODUCTION

Capital market integration has always held a fascination for researchers as it is crucial to understand how two or perhaps more economies can integrate with one another to create a mutually beneficial outcome by eliminating trade boundaries. There has been considerable research on capital market integration, whether domestic or regional, and the results of this research vary from one to the other. For instance, Endri (2009) stated that the degree of integration between ASEAN-5 and China's capital market was still very low, while Suryanta (2011) stated that there was no co-movement whatsoever amongst the ASEAN-5 capital markets.

Santosa's (2017) research stated that the ASEAN-5 countries along with China were significantly integrated to each other, which can be seen as caused by complete integration between China and Singapore, and since Singapore has a significant contribution to the capital market integration in ASEAN-5 countries then it can be concluded that China is integrated with the ASEAN-5 countries. Robiyanto (2019) stated in his research that there is indeed an integration among the ASEAN-5 countries but they were not fully integrated as the Philippines' capital market tend to be segmented from the other capital markets and from the research it was found that 57.09% of principal components caused a co-movement among these countries. Suganda and Hariyono (2018) stated in their research that post–US presidential election there is a long-term and short-term contagion effect among the ASEAN-5 countries created by strengthened trading relations.

2 METHOD

This research uses the daily closing of the composite stock price index among five ASEAN countries namely Indonesia (Jakarta Composite Index of the Indonesia Stock Exchange), Malaysia (Kuala Lumpur Composite Index of the Kuala Lumpur Stock Exchange), Singapore

(Strait Times Index of the Singapore Stock Exchange), Thailand (SET Index of the Stock Exchange of Thailand), and the Philippines (PSEi from the Philippines Stock Exchange), along with China (Shanghai Composite Index of Shanghai Stock Exchange), starting from the period of September 13, 2015, to June 1, 2018. The method used in this research is the orthogonal generalized autoregressive conditional heteroskedastic (OGARCH). The OGARCH model is used to analyze and summarize time series data that is not tied to each other by using principal component analysis (PCA). The OGARCH method has been proven to be the most effective in determining the volatilities and correlation among variables, and has become the most popular to model the conditional covariance in financial time series (Luo, Seco, & Wu, 2015).

Essentially, the OGARCH method would work splendidly on a series of correlated data (Jingjing, 2011). However, first, a correlation analysis is needed to determine whether or not the data used in the study is correlated. After proven to be correlated, then the OGARCH analysis can be conducted. The OGARCH analysis – which is a combination of GARCH that was first introduced by Engle (1982) and PCA (Wold, Esbensen, & Geladi, 1987) – is used to acquire eigenvalue vector and eigenvalue through PCA. The monthly return can be calculated by using the natural logarithm (ln) of the next monthly closing of composite stock price index divided by the monthly closing of composite stock price index, under the assumption that Y is the matrix T X k of the monthly return of composite stock price index at T month.

Since the OGARCH model works best with a series of correlated data, a correlation analysis was conducted to summarize the data's explanatory variation factors before conducting the OGARCH analysis. As can be seen in Table 1, it can be inferred that the data used in this analysis was significantly integrated and suitable to be used in OGARCH analysis.

Based on the OGARCH analysis conducted, it can be seen that the conditional variance of the respective countries' capital market formed two main principal components, as can be seen in Table 2: on the other hand, the eigenvalue and eigenvalue cumulative proportion can be seen in Figure 1.

As can be seen from Table 2, principal component (PC) 1 has an eigenvalue of 2.396 with a proportion of 0.3995, which suggests that the five ASEAN countries and China did not have the same principle variance explanation since it only represents 39.95% of the variance among these countries. It can be inferred that among these five ASEAN countries and China only 39.95% of the conditional variance of the respective countries' capital market return were accounted for. On the other hand, PC 2 resulted in an eigenvalue of 1.005 with a proportion of 0.1675 which implied that the second PC was only able to explain 16.75% of the capital market's variance among these countries. It can be inferred that among these five ASEAN countries along with China that only 16.75% of the conditional variance of these respective countries were accounted for.

Table 1. Correlation of capital market return.

Correlation Prob.	JKSE	KLCI	PSEI	SET	SHCOMP	STI
JKSE	1					
Sig.	-					
KLCI	0.40789	1				
Sig.	0	-				
PSEI	0.418149	0.423722	1			
Sig.	0	0	-			
SET	0.019464	0.008906	0.032018	1		
Sig.	0.5434	0.781	0.3174	-		
SHCOMP	0.151033	0.220928	0.205252	0.017468	1	
Sig.	0	0	0	0.5855	-	
STI	0.389763	0.493872	0.361329	–0.027054	0.327111	1
	0	0	0	0.3983	0	-

Table 2. Result of PCA analysis.

Principal Comp.	Eigenvalue	Proportion	Cumulative Value	Cumulative Prop.
1	2.396837	0.3995	2.396837	0.3995
2	1.005297	0.1675	3.402133	0.567

Table 3. Eigenvectors (loadings).

Variables	PC 1	PC 2
RESID_1_01 (JKSE)	0.453111	0.03456
RESID_2_01 (KLCI)	0.494872	–0.017064
RESID_3_01 (PSEI)	0.459079	0.068687
RESID_4_01 (SET)	0.014332	0.991738
RESID_5_01 (SHCOMP)	0.31024	–0.010896
RESID_6_01 (STI)	0.492529	–0.100666

These two principal components were able to accumulate as much as 56.70% in explaining the conditional variance of these countries' capital market returns and, taking into account the context of capital market integration, this study implied that 56.70% of identical factors could influence the capital market integration among these five ASEAN countries and China. On the other hand, there were 43.30% of accumulated variance that cannot be explained by these two principal components and these factors are indistinguishable and were incapable of affecting the respective countries' capital market concurrently.

3 CONCLUSION

Based on the result of the data analysis conducted in this research, it can be concluded that there is a dynamic co-movement resulting in a significant degree of integration among several ASEAN countries and China: namely, Indonesia represented by Indonesia Stock Exchange (IDX), Malaysia represented by Kuala Lumpur Composite Index (KLCI), the Philippines represented by Philippine Stock Exchange Index (PSEi), Singapore represented by Strait Times Index (STI), and China represented by Shanghai Composite Index (SHCOMP).

Thailand, represented by Stock Exchange Thailand (SET), on the other hand, formed a different principal component. Therefore, it can be inferred that Thailand's stock exchange is more segmented than other ASEAN countries studied in this research. This phenomenon could be a result of internal factors influencing stock movements in Thailand that are more dominant in comparison to external factors in the time period of the study. A notable phenomenon that might have an impact on the segmentation of Thailand's capital market would be the death of Thailand's former king, King Bhumibol the Great, in October of 2016.

For investors or portfolio managers who are thinking of investing in Indonesian Stock Exchange (IDX), Kuala Lumpur Stock Exchange (KLCI), Strait Times Index (STI), Philippines Stock Exchange Index (PSEi), and Shanghai Composite Index (SHCOMP) it is unnecessary for a diversification of the portfolio as the affiliated countries are integrated with one another. Instead, implementing diversification that includes Stock Exchange Thailand (SET) as it is segmented from the other capital markets mentioned earlier, would produce diversification benefit to the investor. Future researchers who are interested in developing a study with the same topic can use longer daily data and include other capital markets from another region such as Europe or America to explore more about inter-regional capital market integration, and to implement another analysis method in conducting the study.

REFERENCES

Bai, J. (2011) "Using orthogonal GARCH to forecast covarince matrix of stock returns." (Doctoral dissertation).

Endri. (2009) "Integrasi Pasar Saham Kawasan Perdagangan Bebas ASEAN-China." *Jurnal Manajemen Bisnis*, 2(2), pp.121–139.

Engle, R. F. (1982) "Autoregressive conditional heteroscedasticity with estimates of the variance of United Kingdom inflation." *Econometrica: Journal of the Econometric Society*, pp. 987–1007.

Luo, C., Seco, L,. and Wu, L. L .B. (2015) "Portfolio optimization in hedge funds by OGARCH and Markov switching model." *Omega*, 57, pp. 34–39.

Robiyanto, R. (2017) "The analysis of capital market integration in ASEAN region by using the OGARCH approach." *Jurnal Keuangan dan Perbankan*, 21(2).

Santosa, B. (2017) "Integrasi Pasar Modal Kawasan Cina-Asean." *Jurnal Ekonomi Pembangunan: Kajian Masalah Ekonomi dan Pembangunan*, 14(1), pp. 78.

Suganda, T. R., and Hariyono, A. R. (2018) "The integration of ASEAN-5 capital market after the Donald Trump election." *Jurnal Keuangan dan Perbankan*, 22(4).

Suryanta, B. (2011) "Capital market integration in ASEAn countries: Special investigation of Indonesian towards the big four." *Asian Journal of Technology Management*, 4(2), pp. 109–114.

Wold, S., Esbensen, K., & Geladi, P. (1987) "Principal component analysis." *Chemometrics and Intelligent Laboratory Systems*, 2(1–3), pp. 37–52.

Facing Global Digital Revolution – Nirmala Arum Janie, Dwi Mulyaningsih & Wahyu Rachmawati (eds)
© 2020 Taylor & Francis Group, London, ISBN 978-0-367-33912-8

Building customer loyalty through customer satisfaction as an intervening variable

D.C. Kuswardani & T.E. Yani
Universitas Semarang, Semarang, Indonesia

ABSTRACT: This research aims at testing service recovery and complaint handling in building customer loyalty through customer satisfaction in transportation rentals. The sample consists of 90 respondents chosen using purposive sampling based on the population, namely, transportation rental users, which amounts to 912 customers. The analysis is done using structural equation modelling (SEM) with AMOS Version: 21 program application. The hypothesis testing results indicate that all hypotheses are confirmed. Service recovery and complaint handling have positive influences on customer satisfaction and customer satisfaction has a positive influence on customer loyalty.

Keywords: service recovery, complaint handling, customer satisfaction, customer loyalty

1 INTRODUCTION

Customer loyalty plays a challenging role for transportation businesses such as car rental service companies. Service recovery, complaint handling, and satisfaction are deemed as the basis of building customer loyalty (Liat et al., 2017). Wahjudi, Kwanda, and Sulis (2018) find that the commitment that accompanies repeat purchase means customers refuse to change and buy the same products/services even when the products or services are rarely found in the market at that moment. Customers voluntarily recommend the products and services to their colleagues, family, or their customers (Venkatesan, 2018). According to Suhail, (2018), customer loyalty is the customers' faithfulness to companies, brands, or products. Farida and Ardyan (2018) suggest that loyalty is the attitude of liking a brand as demonstrated by consistent purchase of that brand at all times. This research tests the influence of recovery service and complaint handling on customer loyalty with customer satisfaction as an intervening variable in transportation rental companies which for the last three years received an increasing number of and more varied complaints.

2 LITERATURE REVIEW & HYPOTHESES FORMULATIONS

Previous studies suggest that service recovery does not waste resources, rather it gives goodwill returns and high profit (Krishna, Dangayach, & Jain, 2011) and is tightly related to customer satisfaction (Kurata & Nam, 2010).
H1: Service recovery has a positive influence on customer satisfaction.

An effective recovery process can fix the service failure and, as a result, can change dissatisfied customers into satisfied ones, improve the relationship with customers, and prevent them from defecting (Nugel dan Santos, 2017). Harris et al. (2006) find that complaint handling has some influence on customer satisfaction.
H2: Complaint handling influences customer satisfaction.

Customer satisfaction and service recovery tend to predict customer intention to make a repeat purchase. The previous research by Baron et al. (2005) finds that satisfaction based on service recovery has a positive impact on customer loyalty.
H3: Service recovery influences customer loyalty through customer satisfaction.

Complaint handling positively influences customer's perception of the organization's professionalism, which might determine an increase in customer satisfaction and end up with customer loyalty (Maxham and Neteneyer, 2002; Nagel, 2017, Ruchtenbeger, 2008).
H4: Complaint handling influences customer loyalty through customer satisfaction.

Based on empirical findings, service quality is one of the antecedents of satisfaction and loyalty is one of the consequences of satisfaction (Cronin & Taylor, 1992, 1994; Dabholkar et al., 2000; Coner & Gungor, 2002). Daniel (2016) found that customer satisfaction influences customer loyalty. Empirical evidence shows that customer satisfaction is translated into loyalty (Erjavec, 2015).
H5: Customer satisfaction has a positive influence on customer loyalty.

3 RESEARCH METHOD

The population in this research is 912 customers of transportation rental companies. The sample is taken using Slovin formula and 90.12 (rounded to 90) samples were obtained. They were taken using purposive sampling technique, using such criteria as having been customers for at least one year and being over 20 years of age.

The validity test with significance value >0.05 means no significant relationship occurs and if the significance value is <0.05, it means a significant relationship occurs. Reliability test is a tool to measure the questionnaire indicators of the variables: if a person's answer is consistent (i.e. variance of p value of variable <0.01 or 1%), then it is said to be reliable (Ghozali, 2016). Normality test aims at testing whether the regression model has a normal distribution or not using graphic analysis (Ghozali, 2016). If c.r kurtosis value >2.58, then the value of univariate test in kurtosis is <2.58, meaning the regression model has a normal distribution. According to Ghozali (2016), hypotheses are tested using a t test, if the probability value ≤ 5%, then H0 is rejected and Ha is confirmed, and if the probability value ≥ 5%, then H0 is confirmed and Ha is rejected.

4 RESULTS & DISCUSSIONS

The validity test uses confirmatory factor analysis (CFA) (Ghozali, 2016). Indicators are said as valid if their significance levels are <0.05, thus all indicators of service recovery, complaint handling, customer satisfaction, and customer loyalty variables are valid, since they produce a significance rate of <0.05. The reliability test generates a composite reliability >0.6, thus it can be concluded that the indicators in all research variables are reliable. The normality test as seen from the normal probability output of the model has overall indicated that the normality assumption was fulfilled. This is because c.r. value of skewness and kurtosis lies between -1.96 and 1.96 (Z at 0.05). The equation of structural equation modelling $Y_1 = 0.177X_1 + 0.837X_2$, meaning (1) as the service recovery improves the customer satisfaction will also improve and (2) as the complaint handling improves the customer satisfaction will also improve.

Influence of *service recovery* on customer satisfaction. Based on the significance value, there is a positive influence of service recovery (0.045) on customer satisfaction. Thus, H1, which states service recovery influences customer satisfaction, is confirmed. This result is supported by Ahmad (2002); Ashraf and Manzoor (2017); Choi et al. (2014). Mueller et al., (2003); Inyang (2015); Ozuem et al. (2017); and Orsingher, Valentini, and de Angelis (2010), who find that service recovery influences customer satisfaction.

Influence of *complaint handling* on customer satisfaction. Based on the significance value, there is a positive influence of complaint handling (0.008) on customer satisfaction. Thus, H2, which states complaint handling influences customer satisfaction, is confirmed. This result is supported by Andreassen (2000); Fornell and Wernerfelt (1987); Svari et al. (2010); Friele, Reitsma, and De Jong (2015); Harris et al. (2006); Orsingher, Valentini, and de Angelis (2010); Rothenberger, Grewal, and Iyer (2008), who find that complaint handling influences customer satisfaction.

Influence of *service recovery* on customer loyalty through customer satisfaction. The influence of service recovery on customer loyalty through customer satisfaction is 0.528 + (0.177 * 0.850) = 0.67845. This shows that service recovery has some influence on customer loyalty through customer satisfaction. This result is supported by Craighead et al. (2004); Harris et al. (2006); Krishna, Dangayach, and Jain (2011); Wu (2016); Assefa (2014); and Yeoh et al. (2014), who find that service recovery influences customer loyalty through customer satisfaction.

Influence of *complaint handling* on customer loyalty through customer satisfaction. The influence of complaint handling on customer loyalty through customer satisfaction is 0.700 + (0.837*0.850) = 1.41145. This means complaint handling influences customer loyalty through customer satisfaction. This result is supported by Kelley et al. (1993); Levesque and McDougall (1996); Callin (2012); Maxham and Neteneyer (2002); Ruchtenbeger (2008) and Nagel (2017), who find that complaint handling influences customer loyalty through customer satisfaction.

Influence of customer satisfaction on customer loyalty. Based on the significance value, it can be said that there is a positive influence of customer satisfaction (0.014) on customer loyalty. Thus, H5 which reads customer satisfaction influences customer loyalty is confirmed. This result is supported by Sivadass and Baker (2000); Bowen and Chen (2001); Ehigie (2006); Faullant, et al. (2008); Baumann, Elliott, and Burton (2012), and Erjavec (2015), who find that customer satisfaction influences customer loyalty.

The determination coefficient (R2) test can be done by seeing the squared multiple correlations at the estimate value of dependent variable (i.e., customer loyalty, Ghozali, 2016). In this research, the obtained value of estimate squared multiple correlations is 0.694, thus it can be interpreted that complaint handling and service recovery variables can explain customer satisfaction at 69.4%. The value of estimate squared multiple correlations is 0.722, hence it can be said that complaint handling, service recovery, and customer satisfaction variables can explain customer loyalty at 72.2%.

5 CONCLUSIONS

Based on the results of analysis and discussion, we concluded that *service recovery* has a positive influence on customer satisfaction, meaning that if service recovery is improved, then customers will be more satisfied. *Complaint handling* has a positive influence on customer satisfaction, meaning that if complaint handling is improved, then customers will be more satisfied. *Service recovery* has a positive influence on customer loyalty through customer satisfaction, meaning that if the service recovery is right and fast, then the satisfied customers will be more loyal. *Complaint handling* has a positive influence on customer loyalty through customer satisfaction, meaning that if complaint handling gets better, customers will be satisfied, and eventually they will be loyal. *Customer satisfaction* has a positive influence on customer loyalty, meaning that if customer satisfaction improves, the customers will be increasingly loyal.

The theoretical implication of this research is that, to build customer loyalty, a transportantion rental company needs to focus on satisfying their customers through service recovery and complaint handling. Service recovery is done by consistently providing immediate response when a failure occurs, and complaint handling should be done by solving the problem resulting from the failure. The managerial implication of this research is that transportantion rental service companies need to make a program related to quick and appropriate service recovery and they need to provide training more frequently to their customer service personnel handling customers' complaints. This research result cannot be generalized in all transportantion rental business types.

REFERENCES

Ashraf, H. A., and Manzoor, N. (2017) "An examination of customer loyalty and customer participation in the service recovery process in the Pakistani hotel industry: A pitch." *Accounting and Management Information Systems*. Bucharest Academy of Economic Studies, 16(1), pp. 199–205.

Assefa, E. S. (2014) "The effects of justice oriented service recovery on customer satisfaction and loyalty in retail banks in Ethiopia." *EMAJ: Emerging Markets Journal*, 4(1), pp. 49–58.

Baron, S., et al. (2005) "Typologies of e-commerce retail failures and recovery strategies." *Journal of Services Marketing*. Emerald Group Publishing Limited.

Baumann, C., Elliott, G., and Burton, S. (2012) "Modeling customer satisfaction and loyalty: survey data versus data mining." *Journal of SeMvices marketing*. Emerald Group Publishing Limited, 26(3), pp. 148–157.

Choi, C. H., et al. (2014) "Testing the stressor–strain–outcome model of customer-related social stressors in predicting emotional exhaustion, customer orientation and service recovery performance." *International Journal of Hospitality Management*. Elsevier, 36, pp. 272–285.

Erjavec, H. S. (2015) "Customer satisfaction and customer loyalty within part-time students." *Journal of Economics and Economic Education Research*. Jordan Whitney Enterprises, Inc, 16 (2),p. 1.

Farida, N., and Ardyan, E. (2018) "The driving of customer loyalty: Relational approach, perceived value and corporate image." *International Journal of Business & Society*, 19(1).

Friele, R. D., Reitsma, P. M., and De Jong, J. D. (2015) "Complaint handling in healthcare: Expectation gaps between physicians and the public; Results of a survey study." *BMC research notes*. BioMed Central, 8(1), p. 529.

Ghozali, I. (2016) "Aplikasi dan Analisis Multivariate dengan proses SPSS." *Semarang: Universitas Diponegoro*.

Harris, K. E., et al. (2006) "Consumer responses to service recovery strategies: The moderating role of online versus offline environment." *Journal of Business Research*. Elsevier, 59(4), pp. 425–431.

Inyang, A. E. (2015) "The buffering effects of salesperson service behaviors on customer loyalty after service failure and recovery." *Journal of Managerial Issues*. JSTOR, pp. 102–119.

Krishna, A., Dangayach, G. S., and Jain, R. (2011) "Service recovery: Literature review and research issues." *Journal of Service Science Research*. Springer, 3(1), p. 71.

Kurata, H., and Nam, S.-H. (2010) "After-sales service competition in a supply chain: Optimization of customer satisfaction level or profit or both?." *International Journal of Production Economics*. Elsevier, 127(1), pp. 136–146.

Liat, C. B., et al. (2017) "Hotel service recovery and service quality: Influences of corporate image and generational differences in the relationship between customer satisfaction and loyalty." *Journal of Global Marketing*. Taylor & Francis, 30(1), pp. 42–51.

Orsingher, C., Valentini, S., and de Angelis, M. (2010) "A meta-analysis of satisfaction with complaint handling in services." *Journal of the Academy of Marketing Science*. Springer, 38(2), pp. 169–186.

Ozuem, W., et al. (2017) "An exploration of consumers' response to online service recovery initiatives." *International Journal of Market Research*. SAGE Publications Sage UK: London, England, 59(1), pp. 97–115.

Rothenberger, S., Grewal, D., and Iyer, G. R. (2008) "Understanding the role of complaint handling on consumer loyalty in service relationships." *Journal of Relationship Marketing*. Taylor & Francis, 7(4), pp. 359–376.

Svari, S., et al. (2010) "A DIP-construct of perceived justice in negative service encounters and complaint handling in the Norwegian tourism industry." *Managing Service Quality: An International Journal*. Emerald Group Publishing Limited, 20(1), pp. 26–45.

Venkatesan, R. (2018) "Measuring customer loyalty with service quality and customer satisfaction in life insurance." *Journal of the Insurance Institute of India*, 5(3), pp. 44–49.

Wahjudi, D., Kwanda, T., and Sulis, R. (2018) "The impact of after-sales service quality on customer satisfaction and customer loyalty of middle-upper class landed housings." *Jurnal Teknik Industri*, 20 (1), pp. 65–72.

Yeoh, P.-L., et al. (2014) "Customer response to service recovery in online shopping." *Journal of Services Research*. Vedatya Institute, 14(2), p. 33.

Facing Global Digital Revolution – Nirmala Arum Janie, Dwi Mulyaningsih & Wahyu Rachmawati (eds)
 ISBN 978-0-367-33912-8

Auditors' perception of the effectiveness level of the red flags method in detecting frauds

Ardiani Ika Sulistyawati, Addien Nurfitriana & Dyah Nirmala Arum Janie
Universitas Semarang, Kota Semarang, Indonesia

ABSTRACT: The purpose of the research is to examine auditor demographic factors such as characteristics (gender, job position, working period), and competence (education, fraud experience, fraud training) on auditors' perception. The respondents are auditors working in KAP (Central Accounting Firm) in Central Java and Jogjakarta. The sample used in the study was 87 from the public accounting firm in the area of Central Java and the general accounting firm in the field of Jogjakarta. Data were collected through questionnaires and processed by multiple linear regression analysis. We utilized SPSS for Windows version 16.0. The results of this study suggest that education and fraud experience influence auditors' perception. Meanwhile, gender, job position, tenure, and fraud auditor training in perceptual not on influence perceptions of red flag effectiveness. Simultaneously, gender variables, job position, education, work period, fraud experience, and fraud training effect auditors' perception.

Keywords: auditors, fraud experience, fraud training, red flags

1 INTRODUCTION

Quality financial reports present information that is true, honest, relevant, reliable, comparable, and understandable. Quality financial statements are relevant because people use them as a basis for decision-making. Besides, high-quality financial report information will erode public confidence in the local government in the presence of various financial scandals. Financial reports that are not high quality (bad) provide opportunities for irregularities and errors in the financial sector, which can lead to fraud, corruption, collusion, and potenitally lawsuits. If it continues, of course, the people will suffer the consequences, and see high transaction costs and poor public services. Furthermore, if this results in a crisis of confidence in both the public and the creditors, international investors will hesitate to place their funds in Indonesia, except with a high burden (Wungow, Lambey, & Pontoh, 2016).

This research is the development of research previously conducted by Rustiarini and Novitasari (2014). The object of this research is external auditors in Central Java and DIY, while the object of previous research was external auditors located in Denpasar Bali. The variables in this study are the same as the previous research variables, namely the independent variables consisting of education, gender, job position, length of employment, experience in detecting fraud, and fraud detection training. Moreover, the dependent variable in this study is the auditor's perception of the effectiveness of the red flags.

2 LITERATURE REVIEW

Indeed, every red flag has a different level of effectiveness in detecting fraud. The difference in assessment is due to differences in perceptions, personal assessment characteristics, incentives, or differences in activities and responsibilities related to work. Even individuals who are in the same profession, namely internal auditors and external auditors, have different perceptions of the effectiveness of red flags (Rustiarini, Suryandari, & Nova, 2016).

Rustiarini and Novitasari (2014) found that female auditors were more likely to detect fraud using red flags than male auditors. Not all auditors are male: many auditors, even at the senior auditor levels, are women. The phenomenon of the average auditors, whether male or female, is one of the reasons gender will influence an auditor in detecting fraud. Women tend to see clients from the emotional side, including body language and clients' nonverbal cues, while men do not pay as much attention to nonverbal cues from clients.

The longer a person holds a position as an internal auditor, the more knowledge and expertise they will possess. The long working period will also provide an opportunity for internal auditors to increasingly explore their own ability to achieve, including increased capacity related to fraud detection. In terms of effectiveness of the red flags in detecting fraud, the longer working period will undoubtedly increase ability to identify the red flags used in detecting fraud (Rustiarini & Novitasari, 2014).

Often the level of formal education someone has is used as a measure of the level of knowledge that person has. To be able to work as an auditor, one must undoubtedly have a specialized education, especially in accounting. The higher the level of the auditor's education, the wider the knowledge possessed, which will affect the ability to make decisions, one of which relates to the determination of the effectiveness of red flags in detecting fraud (Rustiarini et al., 2016). A person who does work according to his knowledge will provide better results than those who do not have sufficient knowledge of their duties. Education will develop essential knowledge, expertise, and discipline in professional audit performance.

Betrika (2015), interpreted experience as the auditor's experience in conducting financial statement audits both in terms of the length of time and the number of assignments. The audit experience will allow a public accountant to become familiar with the situation and circumstances in each assignment. The more auditor experience, the more audit findings can generate suspicion.

In order to increase knowledge, auditors can take formal training, which is considered to have quality equivalent to knowledge from previous experience. The training aims to increase awareness of the potential for fraud within the company. If the auditor has sufficient knowledge and high fraud awareness, sensitivity to the appearance of fraud symptoms will help the auditor effectively identify the red flags appropriate for use in fraud detection procedures (Rustiarini & Novitasari, 2014).

3 RESEARCH METHOD

The population in this study were external auditors who served in the Central Java and DIY KAP. There are 19 KAP spread in Central Java and 11 KAP spread in DIY. The method of determining the sample used is saturated sampling, which is a sampling method that takes all members of the population used as research samples. Data analysis in this study was conducted using multiple linear regression.

4 RESULTS AND DISCUSSION

Questionnaires were distributed by researchers, but because dissemination of the questionnaires occurred in October, which approached the end of the year so many auditors were assigned outside the city, some accounting firms limited the number of questionnaires filled out and some KAPs refused to fill out questionnaires. From the 112 questionnaires distributed 87 or 78% of questionnaires were returned, and 25 or 22% were not returned. Respondents in this study were external auditors who worked for KAP in Central Java and DIY.

This study produced the regression equation model as follows:

$$Y = -0,119\ X_1 + \ 0,014\ X_2 + 0,496\ X_3 - 0,069\ X_4 + 0,370\ X_5 + 0,028X_6 + e$$

Where:
Y = Auditor's perception
α = Constant
β = Regression coefficients
X1 = Gender
X2 = Job position
X3 = Work period
X4 = Pendidikan
X5 = Pengalaman
X6 = Pelatihan
e = error

The adjusted R2 is 0.617. This means that 61.7% of auditors' perceptions of the level of effectiveness of red flags can be explained by independent variables that include gender, work position, education, years of work (tenure), experience, and training. While other variables explain the remaining 38.3% outside this study, the F test results in a significance value of 0.000, which has a value smaller than 5%, which means that in this study it can be concluded that the independent variables simultaneously affect the dependent variable.

This research is in line with research conducted by Baihaqy and Hadri (2012) which states that gender influences auditor perceptions, because women and men get the same knowledge through the development and learning programs about indications of fraud in the company or from science references but get different results.

The results of this study also support the results of the study of Rustiarini and Novitasari (2014), which states that job position does not affect the auditor's perception of the effectiveness of the red flags. They say that, in this case, auditors are required to understand auditing standards when doing the work, especially those related to fraudulent forms of financial reporting. It requires that all auditors have a good understanding of red flags and their use in detecting fraud.

The results of this study support the study of Rustiarini and Novitasari (2014), which states that tenure does not affect the auditor's perception in identifying fraud with the red flags method, but this study contradicts the results of Hegazy and Kassem's (2010) research, which states that years of work have an effect on detection cheating using the red flags method.

Experience affects the auditor's perception of the level of effectiveness in detecting the red flags of fraud. The results of this study support the results of Moyes (2007) and Rustiarini et al. (2016), which state that the auditors' experience in detecting fraud affects their perceptions in assessing the level of effectiveness of red flags. This experience has led to differences in perception between auditors who have more experience and those who have less experience. The greater the number of companies investigated, the more experience the auditor has because the problems or cases that exist in each company are different. Auditors who detect and find fraud in the company have different perceptions from auditors who have never detected or found fraud. It is what causes differences in perception of the level of effectiveness of the red flags method.

The training variable does not affect the auditor's perception of the effectiveness of red flags in detecting fraud. The results of this study contradict the results of Yang et al.'s (2010) research, which states that training has a significant effect on auditor perceptions of the effectiveness of the red flags method in detecting fraud. According to Yang et al., auditors who had attended training and seminars on fraud detection had a better perception of the effectiveness of using the red flags method.

5 CONCLUSIONS

Gender, job position, and working period do not affect the auditor's perception of the effectiveness of the red flags method for detecting fraud. Education influences the auditor's perception of the effectiveness of the red flags method for detecting fraud. Experience affects the auditor's perception of the effectiveness of the red flags method for detecting fraud. Training does not affect the auditor's perception of the effectiveness of the red flags for detecting fraud.

Looking at the results of this study stating that education and experience influence auditor perceptions of the effectiveness of red flags to detect fraud, the researcher suggests that audits should be carried out by auditors who are experienced and have sufficient knowledge, in this case the senior auditor and audit partner. Junior auditors can conduct an audit, but senior auditors must supervise them. This assignment is an effort to maintain the quality and accuracy of the audit results so that the credibility of the audit results in the eyes of users of financial statement information can be maintained and indications of fraud can be detected.

We expect that public accounting firms can further improve education and competence for auditors so that financial statements are following established standards and later the auditor's findings must be accountable and checked by the regulator.

Unfortunately, the actual distribution of questionnaires was not appropriate because it was conducted during a busy period for the accounting firms so this resulted in the majority of firms rejecting and limiting the filling of research questionnaires

REFERENCES

Baihaqy, F., and Hadri, K. (2012) "Persepsi Akuntan Terhadap Indikasi Kecurangan Kecurangan Pelaporan Keuangan." *Jurnal Fakultas Hukum UII*, 16(2).

Hegazy, M., and Kassem, R. (2010) "Fraudulent financial reporting: Do red flags really help?" *Journal of Economics and Engineering*. Progress Publishing, 4, pp. 69–70.

Moyes, G. D. (2007) "The differences in perceived level of fraud-detecting effectiveness of SAS No. 99 red flags between external and internal auditors." *Journal of Business & Economics Research (JBER)*, 5(6).

Rustiarini, N. W., and Novitasari, N. L. G. (2014) "Persepsi Auditor Atas Tingkat Efektivitas Red Flags Untuk Mendeteksi Kecurangan." *Jurnal Akuntansi Multiparadigma*, 5(3), pp. 345–354.

Rustiarini, N. W., Suryandari, N. N. A., and Nova, I. K. S. (2016) "Red flags and fraud prevention on rural banks." *Buletin Ekonomi dan Perbankan*, 19(2), pp. 177–206.

Wungow, J. F., Lambey, L. and Pontoh, W. (2016) "Pengaruh tingkat pendidikan, masa kerja, pelatihan dan jabatan terhadap kualitas laporan keuangan pemerintah Kabupaten Minahasa Selatan." *JURNAL RISET AKUNTANSI DAN AUDITING" GOODWILL"*, 7(2).

Yang, W., et al. (2010) "Professional demographic factors that influence Iranian auditors perceptions of the fraud-detecting effectiveness of red flags." *International Business & Economics Research Journal (IBER)*, 9(1).

Facing Global Digital Revolution – Nirmala Arum Janie,
Dwi Mulyaningsih & Wahyu Rachmawati (eds)

How do companies in Indonesia determine their inventory models?

Abdul Karim, Erika Regina, Yulianti Yulianti & Dyah Nirmala Arum Janie
Universitas Semarang, Kota Semarang, Indonesia

ABSTRACT: We need to use the right inventory method for today's business competition. This research empirically aims at examining the factors influencing the inventory method of the trading and manufacturing companies listed on the Indonesian Stock Exchange. A purposive sampling method was employed in this research with the research population of companies listed on the Indonesian Stock Exchange during the period of 2012–2017: 234 samples were obtained and then analyzed using a logistic regression analytical method. The research results show that company size and supply variability have a significant effect on inventory method selection, while ownership structure and current ratio do not have a significant effect on inventory method selection. The Nagelkerke R square testing result is 29.5%, meaning that those four independent variables explained only 29.5% of the dependent variable, while the other excluded variables explained the remaining 70.5%.

Keywords: company size, ownership structure, current ratio, inventory variability, inventory method

1 INTRODUCTION

The business world has undergone significant changes that greatly impact Indonesian economic conditions, as we can see from the many competing companies and demanding buyers of consumed products and services. With the existence of those many competitors, each company must be able to improve its resource performance so that the company may continue to survive in competition (Riswan & Fasa, 2016).

Earning profits is one way companies survive in tight business world competition. By earning profits, the company owner hopes that the company may continuously grow and develop if the company has good functional management expertise in managing the product processing, marketing, or investment. When obstacles result from product processing, marketing and investment may also be obstructed (Ayem & Harjanta, 2018).

Obstructions may occur in the production process due to several things, including inventory. When obstacles result from the inventory, the production process may also be obstructed. Thus, in obtaining profits, the companies may automatically be obstructed (Ayem & Harjanta, 2018). Inventory management using the correct inventory method may impact the company's sustainability, in that the production process may run smoothly and the customers' needs may be adequately fulfilled (Ayem & Harjanta, 2018).

This research is a replication of research conducted by Nailul Rahmi, Anik Malikah, and Junaidi (2018) on *Analysis of Factors Affecting the Inventory Accounting Method.* This research adds ownership structure as a variable based on the previous research (research gap) showing inconsistent results. This research uses manufacturing and trading companies within the period of 2012–2017, as suggested by the previous researchers to add the companies as the research samples and lengthen the research period.

2 LITERATURE REVIEW

In this research, the use of inventory method is affected by several factors. The research conducted by Syailendra and Raharja (2014) states that company size has a positive and significant effect on inventory valuation method selection. The research conducted by Tjahjono and Chaerulisa (2015), Nugroho Tulus Rahayu and Edy Susanto (2017), and Nailul Rahmi, Anik Malikah and Junaidi (2018) also have similar results as that conducted by Syailendra and Raharja (2014). However, the research conducted by Riswan and Fasa (2016) shows different results in which company size has a positive but insignificant effect on inventory evaluation method selection. The research conducted by Ayem and Harjanta (2018) also shows a negative and insignificant effect on inventory accounting method selection.

According to the research conducted by Syailendra and Prog (2014) and Nugroho Tulus Rahayu and Edy Susanto (2017), ownership structure affects inventory method selection. The research conducted by Syailendra and Raharja (2014) shows that ownership structure negatively and significantly affects inventory method selection while the research conducted by Nugroho Tulus Rahayu and Edy Susanto (2017) shows that ownership structure negatively but insignificantly affects inventory method selection.

The research related to the effect of current ratio on inventory method selection has been conducted by Riswan and Fasa (2016) and Nailul Rahmi, Anik Malikah, and Junaidi (2018). Both researches show different results. Russian and Restiani Fasa (2016) found that current ratio positively but insignificantly affect inventory method selection. However, the research conducted by Nailul Rahmi, Anik Malikah, and Junaidi (2018) shows that current ratio positively and significantly affects inventory method selection.

The effect of inventory variability on inventory method selection has been conducted by Syailendra and Raharja (2014), Ayem and Harjanta (2018), and Nailul Rahmi, Anik Malikah, and Junaidi (2018). The research conducted by Syailendra and Raharja (2014) has concluded inventory variability negatively and significantly affects inventory method selection. Thus, the research conducted by Ayem and Harjanta (2018), and Nailul Rahmi, Anik Malikah, and Junaidi (2018) show similar results.

This research is the development of research conducted by Nailul Rahmi, Anik Malikah, and Junaidi (2018). The research conducted by Nailul Rahmi, Anik Malikah, and Junaidi (2018) used three independent variables: inventory variability, company size, and current ratio. Meanwhile, this research uses four independent variables: company size, ownership structure, current ratio, and stock variability. In addition, Nailul Rahmi, Anik Malikah, and Junaidi (2018) used financial statements of manufacturing companies listed on the Indonesian Stock Exchange during the period of 2014 to 2017, while this research uses the financial statements and trademarks of manufacturing companies listed on Indonesian Stock Exchange between 2012 and 2017.

The reason for adding ownership structure as an independent variable is because there are still inconsistent research results that make this research still interesting to re-examine, especially to find out the inventory method used by most manufacturing and trading companies. By understanding the inventory method used by the manufacturing and trading companies, we may figure out the profits that the company may earn.

3 METHOD

A purposive sampling method was used in this research with the manufacturing and trading companies listed on the Indonesia Stock Exchange during the period of 2012 to 2017 considered as the research population: 234 samples were obtained and then analyzed using a logistic regression analytical method.

4 RESULTS

The testing results show that company size has a significant effect on inventory method with a significance value smaller than 0.05 equal to 0.000. The result of this research is in line with that of research conducted by Syailendra and Raharja (2014), Thajono and Chaerulisa (2015), Rahayu and Susanto (2017), and Rahmi, Malikah and Junaidi (2018). Large companies prefer to use the conventional method. Meanwhile, smaller companies prefer to use FIFO method. By using the FIFO method, the acquisition of profits obtained by the company is larger. The acquisition of profits obtained by the company will also affect tax payments. Companies prefer to use the average method because they may pay lower tax using the average method instead of the FIFO method. Smaller profits indicate that the transfer of wealth out of the company (tax costs) smaller than profits earned using the FIFO method. This is why the companies prefer to use the average method. The research results show that company size has a significant effect on inventory method. This research is not in line with that conducted by Riswan and Fasa (2016) and Ayem and Harjanta (2018). Based on total assets, the larger the company, the higher the possibility of the company using the average method than the FIFO method.

The testing results of this research indicate that ownership structure does not significantly affect inventory method. The significance value obtained in this research is higher than 0.005 equal to 0.258. This is in line with research conducted by Rahayu and Susanto (2017). Based on the testing results, ownership structure does not significantly affect inventory method. Even though managers holding shares cannot resolve conflicts between company owners and managers, company managers must comply with regulations made by company owners because the managers are hired by the company owners. Thus, the managers must obey all decisions made by the company owners, including the decision related to the implementation of inventory method. The company owners have decided to use the average method. Using the average method may result in lower profits that minimize tax payments. Based on the observed data, most companies use the average method. Only three companies consistently use FIFO method. Thus, the tested data are found unbalanced. So the results obtained in the research are not significant. Ownership structure does not significantly affect inventory method. These results are not in line with those of research conducted by Raharja and Syailendra (2014). The result of research conducted by Raharja and Syailendra (2014) shows that ownership structure significantly affects inventory method.

The testing results of this research indicate that current ratio does not significantly affect inventory method with a significance level of greater than 0.05 equal to 0.242. The result of this research is in line with that of research conducted by Riswan and Fasa (2016). The current ratio does not significantly affect inventory method. The company will use a method that generates profits for the company's sustainability. The company does not see the amount of short-term debt that the company uses with the average method. By using this method, the company may see a tax saving. In this research, most companies use the average method; only three companies consistently use FIFO method. Thus, the tested data becomes unbalanced in that the current ratio does not significantly affect inventory method. The result of this research is not in line with that of research conducted by Rahmi, Malikah, and Junaidi (2018), stating that current ratio significantly affects inventory method.

The testing result of this research is that variability significant affects the inventory method with a significance value smaller than 0.05 equal to 0.035. It is in line with the result of research conducted by Syailendra and Raharja (2014), Ayem and Harjanta (2018), and Rahmi, Malikah and Junaidi (2018). This research successfully proves that company variability significantly affects the choice of inventory method. With smaller inventory variation, the profits generated by the company are also small, thus, most companies use average method, and this will allow them to see a tax saving. The companies using the FIFO method have more inventory variations, which means higher profits so the company cannot see a tax saving. Only three companies consistently use FIFO method. Thus, the result of this research shows that inventory variability significantly affects inventory method.

5 CONCLUSION

Most companies considered as the research samples use the average method; only three out of 39 companies used the FIFO method. Thus, the results inadequately explain the dependent variable. The Nagelkerke R square testing result in this research is only 29.5%. It means that the effect of independent variables on the dependent variable is relatively small, by only 29.5%, while other factors affect the remaining 70.5%. The selected variables are considered inadequate. Thus, there are still many other factors which may affect the choice of inventory method.

These research results and limitations may provide better input for further researchers. We should add the research samples, not only the manufacturing companies and trade, so that the results could get better and could explain the dependent variable. Due to the Nagelkerke R square testing result which is only 29.5%, we expect that the further research may add more testing variables as intervening variables for in-depth analysis, which may affect the choice of inventory method.

REFERENCES

Ayem, S., and Harjanta, A.P.P. (2018) "Pengaruh Ukuran Perusahaan, Variabilitas Persediaan, Kepemilikan Manajerial, Financial Leverage dan Laba Sebelum Pajak Terhadap Pemilihan Metode Akuntansi Persediaan (Studi Empiris Pada Perusahaan Sub Sektor Farmasi yang terdaftar di Bursa Efek Indonesia Periode 2012-2016)." *Akuntansi Dewantara*, 2(1), pp. 83–95.

Riswan, and Fasa, R. (2016) "Analisis Faktor-Faktor yang Mempengaruhi Pemilihan Metode Penilaian Persediaan Pada Perusahaan Dagang yang Terdaftar di Bursa Efek Indonesia Periode 2010-2014." *Jurnal Akuntansi & Keuangan*, 7(2), pp. 193–204.

Syailendra, B., and Raharja, R. (2014) "*Analisis Faktor-Faktor yang Berpengaruh Terhadap Pemilihan Metode Penilaian Persediaan (Studi Kasus Pada Perusahaan Dagang dan Manufaktur yang Terdaftar di Bei Tahun 2008-2012)*" Doctoral dissertation, Fakultas Ekonomika dan Bisnis.

Tjahjono, A., and Chaerulisa, V. N. (2015) "Analisis Faktor-Faktor Yang Berpengaruh Terhadap Pemilihan Metode Akuntansi Persediaan Pada Perusahaan Sub Sektor Perdagangan Besar Barang Produksi Dan Sub Sektor Perdagangan Eceran yang Terdaftar di Bursa Efek Indonesia (BEI)." *Kajian Bisnis STIE Widya Wiwaha*, 23(2), pp. 150–161.

Facing Global Digital Revolution – Nirmala Arum Janie, Dwi Mulyaningsih & Wahyu Rachmawati (eds)
© 2020 Taylor & Francis Group, London, ISBN 978-0-367-33912-8

The income-smoothing model moderated by industry type in Indonesian manufacturing companies in 2014–2017

Yulianti Yulianti, Nyimas Siti Mukarahma Rahim, Abdul Karim &
Dyah Nirmala Arum Janie
Universitas Semarang, Kota Semarang, Indonesia

ABSTRACT: A steady increase in profits will provide a sense of security in terms of investment. The management strategy commonly used at all times is income smoothing. The information presented in financial statements is a consideration for management carrying out income-smoothing practices. This investigation intends to determine the effect of profitability and company size on income-smoothing practice with industry type as a moderating variable. The type of data in this study includes quantitative data and secondary data obtained from the financial statements of each company. The population in this study is companies listed on the Indonesia Stock Exchange from 2014 to 2017. The sampling method uses a purposive sampling method. Selected samples according to criteria amounted to 63 companies from the financial and manufacturing sectors. The data analysis technique uses logistic regression analysis with interaction tests, or moderated regression analysis (MRA), because the dependent variable is a dummy variable, and there are moderating variables. Results of this study indicate that (1) profitability does not affect the practice of income smoothing; (2) company size influences the practice of income smoothing; (3) type of industry is not able to moderate profitability against income smoothing practices; and (4) type of industry can moderate the size of the company against the practice of income smoothing.

Keywords: profitability, company size, income smoothing, industry type

1 INTRODUCTION

The profit obtained from the company's performance in managing its assets is known through management's financial statements of each company. The company can show the results of its work from financial statements that contain company financial information, which will be useful for users of financial information in the decision-making process. Investors will pay more attention to the level of company stocks whose profits are stable than those of companies with unstable profit levels. Company performance is said to be good if the company's shares show a stable profit level (Tuty & Indrawati, 2007, in Sujana, 2014).

Investors, in assessing a company, see how the company's management can utilize the company's assets in obtaining profits (Sidartha & Erawati, 2017). Investors rarely pay attention to the process of profit generated, which ultimately results in company managers doing deviations such as dysfunctional behaviour (Dewi & Prasetiono, 2012, in Sujana, 2014).

So far, there are some phenomena regarding income smoothing that occurs in various companies in Indonesia. One of them is the income-smoothing case that occurred at PT Saratoga Investama Sedaya Tbk (SRTG), where, during the first semester of 2016, the company's investment portfolio increased to 26% on December 31, 2015, from the previous figure of Rp. 13.6 trillion and increased to IDR 17.1 trillion. Increasing market value for corporate investments in various sectors also increases the portfolio. SRTG adopted Statement of Financial Accounting Standards (SFAS) 65, except for consolidation in the statement of financial performance in the first half of 2016. IAS 65 is likely to make SRTG start the application of the fair value of assets of the investment. The implementation of this change will apply in the future and will affect the report on the financial performance of each company in 2016, compared to the consolidated financial statements in 2015 (Setiawan, 2016).

The occurrence of income smoothing is due to incomplete information between the management and the owner of the company. Management has more information related to the company because management is directly involved in the company's operational process while the owner will only have information about the company presented in the financial performance report. Income-smoothing practices allow management to increase stock sales, increase tax rates, and get bonuses (Arum, Nazar, & Aminah, 2017).

This study refers to the research gap or inconsistencies or differences in the results of previous studies. We can find differences in results in the variable profitability and size of the company. Associated with the moderating variable taken by researchers is industry type, such as manufacturing companies and the financial sector, and these have different profits. This research is interested in using the objects of manufacturing companies and the financial sector listed on the Indonesia Stock Exchange (IDX) for the period 2014 to 2017 (IDX, 2019).

2 LITERATURE REVIEW

It is the consideration of investors in investing their capital in companies of interest, namely by looking at and assessing profitability aspects, because investors believe good profitability will result in higher profits. However, if the profitability is not suitable, it will make the management worry because then it will cause investor confidence to decline in the company. That way if the company's profitability is low, it will increase the possibility of company managers to practice income smoothing. ROA (return on asset) tool that measures in this profitability variable. Research by Sari and Amanah (2017) and Sujana (2014) give results that profitability has a positive and significant influence on income smoothing practices. However, researchers (Arum et al., 2017) explain that profitability has a negative and significant effect on income smoothing practices.

H1: Allegedly, profitability affects the income smoothing practice.

Company size influences the structure of funding in the operations of company activities, because a large company is bound to need more funds to run the activities of the company. A large company certainly wants high-profit growth because it can make it easier for companies to obtain loans from creditors. Higher total assets will increase the likelihood of managers practicing income smoothing because creditors do not like unstable earnings growth. Research conducted Fitriani (2018) states that firm size has a significant positive effect on income-smoothing practices, while the results of research Iskandar and Suardana (2016) explain that firm size has a significant negative effect on income smoothing practices.

H2: Alleged size of company influences the practice of income smoothing.

Public industries listed on the Indonesia stock exchange (BEI) have differences related to their financial statements, in terms of profits generated by the company, and how these profits affect the profitability of the company. High profitability means showing excellent performance to obtain profits from the operations of the company, which later also affects investors in making decisions to invest their shares in the company. This is evidenced by Sujana's (2014) research, which states that profitability has a positive effect but is not significant in the practice of income smoothing with industry type moderation.

H3: Allegedly the type of industry moderates the profitability of income smoothing practices

All industries from the manufacturing sector or financial sector must have information relating to different profits as well. From this, we can tell that if the size of the company in each industry is proliferating, the company's financial statements must be good too, so managers will consider profit manipulation to attract investors. Research conducted by Sidartha and Erawati (2017) states that the size of the company has a negative effect but is significant in the practice of income smoothing with moderation in the type of industry.

H4: Allegedly the type of industry moderates the size of the company against the practice of income smoothing

3 METHOD

This study included quantitative data and secondary data obtained from the financial statements of each company. The population in this study are companies listed on the Indonesia Stock Exchange from 2014 to 2017. The sampling method uses a purposive sampling method. Selected samples according to criteria amounted to 63 companies from the financial sector and manufacturing sector. The data analysis technique uses logistic regression analysis with interaction tests or called moderated regression analysis (MRA).

4 RESULTS

The results of the first hypothesis test conducted by logistic regression measures the variable profitability and shows return on assets (ROA) with a coefficient of -1.683 and a significance level of 0.419. We can say that profitability does not affect the income-smoothing practice, and this study rejects H1. These results support the results of studies that have been previously examined by Fitriani (2018) who found that profitability does not affect the income-smoothing practice. Profitability is the company's ability to earn profits. If the profitability of a company increases, the profits obtained by the company also increase. This increase in profits will tend to increase the number of company assets. Good and increasing profitability will also benefit investors because of the company's efforts to manage invested funds to produce the expected return on investors.

The results of the second hypothesis test show the company size variable with the coefficient value of -0.692 and the significance level of 0.002, thus smaller than alpha (0.05). From these results, we can say that the size of the company influences the practice of income smoothing, and this study accepts H2. These results support research by Arum, Nazar, and Aminah (2017), Sidartha and Erawati (2017), and Iskandar and Suardana (2016) that state that firm size influences the practice of income smoothing. The size of the company itself is calculated based on number of employees and total assets of the company. The measurement of the previous results is that the size of the company is calculated using a sales log so that the size of each company is known.

The third hypothesis test results are also carried out by logistic regression with the interaction between profitability and the type of industry. The coefficient of interaction gets a positive value of 6.132, and the significance level of 0.351 is more significant than alpha 0.05. We can say that type of industry is not able to moderate the influence of profitability on income-smoothing practices. Hence it is concluded that we reject H3. These results are supported by research conducted by Sujana (2014), which obtained results that the moderating variables of the industry were not able to strengthen the relationship of profitability to income-smoothing practices. Type of industry is a business field developed by each company so that the company's industry can be known to everyone, and makes investors interested and participate in collaboration in the business. Profitability is a consideration of investors in making decisions to invest their funds. However, according to the results of this study, type of industry is not able to strengthen the relationship of profitability to income-smoothing practices.

The results of the fourth hypothesis test also have an interaction variable between firm size variables and industrial type variables resulting in a coefficient with a positive value of 1.168 and a significance level of 0.003, thus smaller than alpha (0.05). We can assume that type of industry can moderate or strengthen the influence of firm size on income-smoothing practices, and then state that in this study, H4 is accepted. The results of this test are not the same as the results of previous studies, namely Sidartha and Erawati (2017), which obtained results that the moderating industry has a negative and significant effect on income-smoothing practices, while the results of Sujana's (2014) researcher suggest industry type moderating variables have a negative but not significant effect on income smoothing practices. The larger the size of the company, the more the company's assets will draw the attention of investors and the government. A large company is considered to have the ability to obtain high profits so that investors are more confident to invest in the company.

5 CONCLUSION

The independent variable in the study is still limited in explaining the dependent variable. It appears Nagelkerke's R-square value is 0.102 or equal to 10.2%. Future researchers may continue other independent variables thought to influence income-smoothing practices. Also, researchers incorporate other independent variables that influence the practice of income smoothing, such as taxes, stock quotes, company age, managerial ownership, compensation bonuses, dividend policy, debt contracts, and so forth. We can also add or expand the type of industry and extend the observation period.

REFERENCES

Arum, H. N., Nazar, M. R., and Aminah, W. (2017) "Profitabilitas, Ukuran Perusahaan, dan Nilai Perusahaan terhadap Praktik Perataan Laba." *Jurnal Riset Akuntansi Kontemporer*, 9(2), pp. 71–78.

Fitriani, A. (2018) "Pengaruh Profitabilitas, Ukuran Perusahaan, dan Financial Leverage terhadap Praktik Perataan Laba (Income Smoothing) pada Perusahaan Farmasi yang Terdaftar di Bursa Efek Indonesia Periode 2011-2015." *Jurnal Samudra Ekonomi dan Bisnis*, 9(1), pp. 50–59.

IDX. (2019) *Laporan Keuangan dan Tahunan, Bursa Efek Indonesia.*

Iskandar, A. F., and Suardana, K. A. (2016) "Pengaruh Ukuran Perusahaan, Return on Asset, dan Winner/Loser Stock Terhadap Praktik Perataan Laba." *E-Jurnal Akuntansi*, pp. 805–834.

Sari, I. P., and Amanah, L. (2017) "Faktor Faktor yang Mempengaruhi Income Smoothing pada Perusahaan Manufaktur di BEI." *Jurnal Ilmu dan Riset Akuntansi*, 6, pp. 1–19.

Setiawan, D. (2016) *Portofolio Investasi SRTG tumbuh 26%*, https://investasi.kontan.co.id/.

Sidartha, A. R. M., and Erawati, N. M. A. (2017) "Pengaruh ukuran perusahaan dan risiko keuangan pada praktik perataan laba dengan variabel pemoderasi jenis industri." *E-Jurnal Akuntansi*, pp. 1103–1132.

Sujana, I. K. (2014) "Pengaruh Ukuran Perusahaan dan Profitabilitas pada Praktik Perataan Laba dengan Jenis Industri sebagai Variabel Pemoderasi di Bursa Efek Indonesia." *E-Jurnal Akuntansi*, pp. 170–184.

Facing Global Digital Revolution – Nirmala Arum Janie, Dwi Mulyaningsih & Wahyu Rachmawati (eds)
© 2020 Taylor & Francis Group, London, ISBN 978-0-367-33912-8

Tourist consumer behavior patterns in Indonesia: A conceptual framework based on tourism behavior paradigm

Suhendrojono Christantius Christantius Dwiatmadja, John J.O.I. Ihalauw & Apriyani Dorkas
Universitas Kristen Satya Wacana, Salatiga, Indonesia

ABSTRACT: Tourism behavior is a concept and approach to see the system that develops in the world of tourism itself. Tourist consumer behavior is complex and dynamic. Changes in human behavior and society impacts perceptions, decision-making choices, and interest in something. The purpose of the research is to analyze and explore aspects of tourist consumer behavior patterns. We can see the mindset of tourist consumers in Indonesia through the paradigm of tourism behavior. We based this research on observation, using a case study in Yogyakarta Special Region. The method of this research uses a quantitative approach through survey and the analysis of secondary data. The result of this study indicates that tourist consumer behavior has a variety of fundamental aspects. There also various influences on tourist consumer behavior. We can demonstrate aspects of consumer tourism behavior through tourist income, lifestyle, purchasing power, tourism products offered, and the marketing models offered. Higher income, and increased tourism products and marketing models will increase interest in visiting the tourist destination in Yogyakarta.

Keywords: tourist consumer behavior, tourism behavior, tourist destination

1 INTRODUCTION

The development of tourist travel has been increasing. The tourism industry and travel has an impact on numerous activities that we consider a simultaneous part of tourism behavior. Studies on tourism consumer behavior also have received attention in recent years. Social and community change highlights the importance of consumer-behavior studies in tourism since the change and development of people's behavior can influence intention, motivation, and preference of a tourist destination. McColl, Jr, and colleagues (1994) define consumer behavior as "the activity a person takes towards searching and using products and services, including the decision-making process that precedes and determines the activity," in other words, the actions that bring tourists to choose a tourist destination. These actions are crucial for the development of the tourism industry. Other studies showed that tourists are the products of changing population demographics and are more experienced, flexible, and independent minded. Tourists have changed values and lifestyles, which affect tourist demand (Poon, 1993). Comparing Asian and American tourists in terms of travel intention shows that Asian people are now looking for meaningful holidays and prefer a rural experience, while American tourists look for cultural and educational meaning (Swarbrooke & Horner, 1999).

Indonesia is a country that has abundant natural and cultural wealth, leading to it being called a land of "heaven," as in the old saying that wooden and stone sticks can be plants. In Javanese philosophy, Indonesia is a country whose *gemah ripah loh jinawi, tata tentrem kerta raharja*. It means, Indonesia is a culturally rich country, though the beauty of nature and the diversity of the vibrant and numerous Indonesian culture has not been utilized for tourism activities to improve the welfare of the Indonesian people. Tourism has become a concern for

Indonesian government policy: specifically, studies on consumer behavior in tourism have become the central area of studies in tourism, travel, and industry. According to Vukonić (2012), modern tourism comes tourists' passion for a particular destination. Tourists make a trip to specific destinations in large numbers, for personal interest and personal reasons (Vukonic, 2012). In terms of material and spiritual considerations, tourism development depends on the readiness of those dealing with tourists; tourists' perception of the destination upon arrival; the organization of travel toward their destination; and accommodations, especially as it relates to facilities at the destination (Mihajlović and Koncul, 2016).

Indonesian society has long had a tourist orientation, and the traveling behavior of Indonesian people is considered high. We can see this phenomenon from the study-tour activities carried out by educational institutions ranging from the level of kindergarten to higher education. The study tour is one form of travel behavior by Indonesian people known in the world of education. In society, the tradition carried out by a particular group of people is making a pilgrimage to a place that is considered sacred and has philosophical values, and this behavior has been around for a long time. "Comparative study," to gain knowledge and insights about science and the fields of competence possessed by state administrators in strengthening public services, is inseparable from traveling activities.

We rarely find Indonesian people traveling or making unique and varied tourist trips to other countries. Therefore, tourism has a broad definition, not only as a journey but also as a change and movement with cultural, social, economic, and psychological elements. The Indonesian state has many different types of tourism that do not exist in other countries, such as traveling with study tours conducted by educational institutions. Another kind of traveling behavior involves visiting holy and historical places in the form of pilgrimage, for example, a pilgrimage to the tomb of Wali Sanga, which spreads in various regions. There is also traveling with a picnic, as state institutions usually, when conducting official visits while traveling. Yet another kind of travel behavior called "going home," usually undertaken by migrants who work or live in other areas outside their home village. On certain days or events, they take a trip home while on vacation and travel to places that were previously visited or never visited with family. Usually, the phenomenon of "going home" and traveling is carried out on religious holidays and New Years'. The phenomena cause a change in tourist consumer behavior and influence tourism behavior. Mihajlović and Koncul (2016) underline that changes in tourism have some criteria, most notably, the length of the tourist's stay, though also travel organization to a destination, the means of transport, and the area where tourism takes place. We consider specific attractive factors as crucial factors influencing the selection of destination. Moreover, we consider the season and the distance from place of residence (Mihajlović & Koncul, 2016).

Indonesian people know about the life of traveling. It is not something new or foreign to Indonesian society. We can see their travel behavior from three aspects: 1). Historical aspects, where Indonesian people who had left their hometown are finally longing to go home and enjoy their lives while traveling. 2). Cultural aspects, the life of a dynamic and cultured society which tends to see and feel something new and is curious about the atmosphere of life of other people, so the community travels. For instance, several areas such as Bali, Yogyakarta, Solo, Central Java, and other places that have high cultural values are popular sites to visit. 3). Religious aspects and traditional religious relations in some communities have had an impact on the community to travel to get inner and spiritual satisfaction through religious tourism on specific months or days. The term pilgrimage is part of a religious ritual to build a system of belief and values between religious and cultural traditions, which can be united to have an impact on tourism behavior.

Yogyakarta has a tourism development direction and policy based on DIY Regional Regulation No. 1 of 2012, concerning the DIY Tourism Development Master Plan (RIPPARDA). It explains that the direction of the development of DIY tourism is culturally sound. The vision of DIY Tourism in 2025 is to be the leading tourist destination in Southeast Asia (Statistika Pariwisata DIY, 2007). The purpose of this research focuses on tourist consumer behavior patterns in Indonesia through a case study in Yogyakarta, in which we see the conceptual framework of the tourism-behavior paradigm that emphasizes the tourism system.

Research method in this study employs quantitative and qualitative analysis. The qualitative analysis was based on the researcher experience and empirical observation and literature studies. The research gap in this research shows that previous research is not using the approach of a comprehensive system to understand tourism behavior.

2 RESULT AND DISCUSSION

Yogyakarta is a unique region for tourists, and this is why it is a popular tourist destination. Another factor is simply because this city has a long history and is memorable after people visit. Yogyakarta's palace and the people of this city impress tourists who visit and they are with left good memories and unforgettable experiences. Through tourist behavior patterns, we can learn and experience the impression and expression of tourists when they visit this tourist destination. Tourist consumer behavior becomes a significant point in an attempt to understand and analyze the pattern and forms of tourist behavior in the context of the tourism system. Tourist behavior is also a useful indicator for successful development of a tourism agenda. Therefore, an understanding and knowledge of tourist behavior is a fundamental element in the development of tourism, knowing the behavior of tourists has practical value for all tourism stakeholders (Pearce, 2005). Erasmus, Boshoff, and Rousseau (2001) consider that the study of consumer behavior must adapt to the specific situations or products that are the subject of purchase. Individual decisions in the decision-making process can be more or less risky, depending on the final product. Their model of consumer behavior also includes all the steps that occur well before the purchase and afterward. In general, the primary purpose of the consumer behavior models is to provide a simplistic picture of the factors that drive consumers' buying decision-making process (Swarbrooke & Horner, 2007). Seaton (1994) recognizes the intangible nature of tourism services as being the crucial reason explaining why we should pay special attention to the tourist decision-making process. Kotler, Bowen, and Makens (1999) state that globalization has changed tourist consumer behavior, impacting cultural criteria (culture, subculture, and social class), social criteria (reference groups, family, roles, and status), personal criteria (age and life cycle stage, occupation, economic circumstances, lifestyle, personality, and self-concept), and, psychological criteria (motivation, perception, learning, beliefs, and attitudes). In the tourism industry, the quality of human resources tourism has an essential role, becoming a bridge, communicator, and catalyst of the tourism marketing product. The process begins, essentially, with the quality of human resources, which can form the character of tourism.

Yogyakarta, based on empirical observation and study, shows that memory and emotional, and psychological aspects have a substantial impact on tourist consumer behavior patterns, which mostly consist of cultural, social, personal, and psychological factors, directly influencing tourism behavior. Product behavior influences the tendency of tourism behavior in visiting tourist destinations. The category of product behavior referred to in this study consists of essential elements such as attractions, accessibility, amenities, and networks. Besides, several marketing behavior dimensions influence tourism behavior. Tourist consumer behavior in this study consists of products, prices, places, and promotions, which shows that marketing behavior influences tourism behavior. The direction of the influence is positive, meaning that increased marketing behavior could have an impact on increasing tourist visits.

3 CONCLUSION

The results of this study indicate that tourist consumer behavior has a strategic role in building interest in tourist visits. Tourist consumer behavior includes behavior making decisions, encouraging motivation, and creating choices about designated tourist destinations. Tourism behavior as a system and paradigm to see the development and dynamics of tourist consumer behavior shows meaningful relationships of influence in formulating tourism products, tourism marketing, and tourist behavior in all aspects and dynamics of tourism. Understanding tourist consumer behavior becomes an integral part of tourism research and research in an area or country. The consumer behavior pattern shows that the better the impression the tourist consumer has toward the destination and the impression of the society, the better the intensity and interest of tourists to visit.

REFERENCES

Erasmus, A. C., Boshoff, E., and Rousseau, G. G. (2001) "Consumer decision-making models within the discipline of consumer science: A critical approach." *Journal of Consumer Sciences*. South African Association for Family Ecology and Consumer Sciences, 29 (1).

Kotler, P., Bowen, J., and Makens, J. (1999) *Marketing for Hospitality and Tourism*. 2nd ed. USA: Prentice-Hall.

McColl Jr, K. et al. (1994) *Marketing: Concepts and Strategies*. Acumen Overseas Pte. Ltd, Singapore.

Mihajlović, I., and Koncul, N. (2016) "Changes in consumer behavior: The challenges for providers of tourist services in the destination." *Economic Research-Ekonomska istraživanja*. Sveučilište Jurja Dobrile u Puli, Odjel za ekonomiju I turizam'Dr. Mijo . . ., 29(1), pp. 914–937.

Pearce, P. L. (2005) *Tourist Behavior: Themes and Conceptual Schemes*. Channel View Publications.

Poon, A. (1993). *Tourism, Technology, and Competitive Strategies*. CAB international.

Seaton, A. V. (1994) "Tourism and the media," in Witt, S. J., and Miuntinho, L. (eds), *Tourism Marketing and Management Handbook*. Prentice-Hall.

Statistika Pariwisata DIY. (2007) *Statistika Pariwisata DIY*. Yogyakarta: Dinas Pariwisata DIY.

Swarbrooke, J., and Horner, S. (1999) *Consumer Behavior in Tourism*. Great Britain: Butterworth-Heinemann.

Swarbrooke, J., and Horner, S. (2007) *Consumer Behavior in Tourism*. 2nd ed. Burlington, USA: Elsevier.

Vukonic, B. (2012) "An outline of the history of tourism theory: Source material (for future research): Boris Vukonić." in *The Routledge Handbook of Tourism Research*. Routledge, pp. 35–59.

Facing Global Digital Revolution – Nirmala Arum Janie, Dwi Mulyaningsih & Wahyu Rachmawati (eds)
© 2020 Taylor & Francis Group, London, ISBN 978-0-367-33912-8

Indonesian food and beverages companies' values: What affects them?

Dian Indriana Tri Lestari, Destania Ardela Dwiana Pertiwi, Nisetyo Wahdi & Dyah Nirmala Arum Janie
Universitas Semarang, Kota Semarang, Indonesia

ABSTRACT: This study aims to analyze the effect of dividend policy, debt policy, and company size on firm value in the food and beverages subsector of manufacturing companies listed on the Indonesia Stock Exchange. The sampling method used in this study is the purposive sampling method, which is a sampling method based on specific criteria. Samples that meet the criteria in this study are 35 data manufacturing companies in the food and beverages subsector on the Indonesia Stock Exchange in the period 2013–2017. The analysis used is multiple regression analysis that serves to see the effect of the independent variables on the dependent variable, which is preceded by the classic assumption test that consists of the normality test, multicollinearity test, autocorrelation test, and heteroscedasticity test. While hypothesis testing is done using the F test and t-test. The results of data analysis or regression results indicate that partially dividend policy, debt policy, and company size influence the value of the company. Simultaneously dividend policy, debt policy and company size contribute influence of 72.8% to the level of the dependent variable, namely the value of the company.

Keywords: dividend policy, debt policy, company size, company value

1 INTRODUCTION

Dividend policy is a decision that relates to determining whether company profits will be distributed to shareholders as dividends or will be held as retained earnings and subsequently reinvested in the future (Suhartono, 2004). Dividend policy is measured using a dividend payout ratio. According to Sasti, Rina Tjandrakirana, and Ilham (2015), dividend policy affects the value of the company, while research conducted by Normayanti (2017) states that dividend policy does not affect the value of the company.

The ratio of financial leverage measures how far the company's assets are financed by debt. Knowing the financial leverage ratio, we can assess the company's obligations to other parties, the company's ability to fulfil fixed obligations, and the balance between the value of fixed assets and capital. To measure it we used debt to assets ratio. Research conducted by Desi Irayanti (2014) states that debt policy affects the value of the company. However, the research conducted by Sasti, Rina Tjandrakirana and Ilham (2015), and Normayanti (2017) states that debt policy does not affect the value of the company.

Company size is one indicator for measuring company performance. Large company size can reflect if the company has a high commitment to improving its performance continuously, so the market will want to pay more to get its shares because they believe they will get a welcome return from the company. The research conducted by Widiastari and Yasa (2018) states that the size of the company influences the value of the company. However, research conducted by Pantow, Murni and Trang (2015) states that the size of the company does not affect the value of the company.

Based on the research gap above, formulating financial performance factors thought to influence the value of the company, we studyied dividend policy, debt policy, and company size. In the next section, the factors that influence the value of the company selected as independent variables in this study, and the value of the company becomes the dependent variable.

2 LITERATURE REVIEW

Signaling theory emphasizes that dividend payments are a signal for investors that the company has the opportunity to grow in the future, so dividend payments will increase market appreciation for the shares of companies that distribute these dividends. Thus dividend payments have positive implications for company value. Dividend policy is an integral part of the company's funding decisions. According to Lease et al. in Tatang Ary Gumanti (2003) "interpreting dividend policy is a practice carried out by management in deciding to pay dividends, which includes the amount of the rupiah, the pattern of cash distribution to shareholders." The dividend payout ratio determines the amount of profit that we can hold as a funding source. The higher the retained earnings, the less the amount of profit allocated to pay dividends (Cashier, 2010). If the large dividends are distributed, this will increase the share price, which also increases the value of the company. According to Sasti, Rina Tjandrakirana, and Ilham (2015), dividend policy affects the value of the company.

H1: There is a significant influence between dividend policy and company value.

According to Kasmir (2008), the solvency ratio or leverage ratio is used to measure the extent to which a company's assets are financed by debt. Increasing the level of leverage means the level of uncertainty of the return that the company owner will get will be higher too, but at the same time, it will also increase the amount of return obtained. Variable leverage ratio measures the extent to which companies use funding through debt. The higher the level of company leverage, the greater the amount of debt used and the higher the business risk faced by the company if the company's condition deteriorates. Research conducted by Desi Irayanti (2014) states that debt policy affects the value of the company.

H2: There is a significant influence between debt policy and company value.

Company size is one indicator measuring a company's performance. Large company size can reflect if the company has a high commitment to improving its performance continuously, so the market will want to pay more to get its shares because they believe they will get a welcome return from the company. The larger the size of the company, the higher the investor's confidence in the company's ability to provide a return on investment (Sofyaningsih and Hardiningsih, 2011). Large companies tend to have better operating results so they are more able to provide a return on investment that is more profitable than smaller companies. The results of research conducted by Widiastari and Yasa (2018) state that the size of the company influences the value of the company.

H3: There is a significant influence between company size and company value.

3 METHOD

The sampling method used in this study is the purposive sampling method, which is a sampling method based on specific criteria. Samples that meet the criteria in this study are 35 data manufacturing companies in the food and beverages sub-sector on the Indonesia Stock Exchange in the period 2013–2017. The analysis used is multiple regression analysis which serves to see the effect of the independent variables on the dependent variable, which is preceded by the classic assumption test which consists of the normality test, multicollinearity test, autocorrelation test, and heteroscedasticity test, while hypothesis testing is done using the F-test and t-test.

4 RESULTS

This study produced a coefficient of determination of 72.8%. This shows that the independent variables, namely dividend policy, debt policy, and company size, contribute an influence of 72.8% to the level of the dependent variable, namely corporate value. While the remaining 27.2% is influenced by other variables, we did not mention them in this study. The F value shows the results of the statistical test with a significance of 0.000. Provided that the significance value is less than 0.05, it means that there is a significant effect of the independent variables together on the dependent variable. In other words, dividend policy, debt policy, and company size jointly influence the value of the company.

This study shows that dividend policy has a significant effect on firm value. Signaling theory emphasizes that dividend payments are a signal for investors that the company has the opportunity to grow in the future, so dividend payments will increase market appreciation for the shares of companies that distribute these dividends. Thus dividend payments have positive implications for company value. Dividend policy is an integral part of the company's funding decisions. According to Lease et al. in Tatang Ary Gumanti (2003) "interpreting dividend policy is a practice carried out by management in deciding to pay dividends, which includes the amount of the rupiah, the pattern of cash distribution to shareholders." The dividend payout ratio determines the amount of profit that we can hold as a funding source. The higher the retained earnings, the less the amount of profit allocated to pay dividends (Cashier, 2010). If large dividends are distributed, this will increase the share price, which also increases the value of the company. The results of the study follow Sasti, Rina Tjandrakirana, and Ilham (2015), which states that dividend policy affects the value of the company.

This research proves that debt policy has a significant effect on firm value. According to Kasmir (2008), the solvency ratio or leverage ratio measures the extent to which a company's assets are financed by debt. Increasing the level of leverage means the level of uncertainty of the return that the company owner will get will be higher too, but at the same time, it will also increase the amount of return obtained. Variable leverage ratio measures the extent to which companies use funding through debt. The higher the level of company leverage, the greater the amount of debt used and the higher the business risk faced by the company if the company's condition deteriorates. Research conducted by Desi Irayanti (2014) states that debt policy affects the value of the company.

This research proves that company size influences company value. Company size is one indicator measuring a company's performance. Large company size can reflect if the company has a high commitment to improving its performance continuously, so the market will pay more to get its shares because they believe they will get a welcome return from the company. The larger the size of the company, the higher the investor's confidence in the company's ability to provide a return on investment (Sofyaningsih and Hardiningsih, 2011). Large companies tend to provide more exceptional operating results so that they have a greater ability to provide a return on investment that is more profitable than smaller companies. Widiastari and Yasa (2018) stated that the size of the company influences the value of the company.

5 CONCLUSION

The results of this study prove that dividend policy and debt policy have an effect on company value, and company size influences the value of food and beverage companies in Indonesia during 2013–2017. In future, we can test other cash position factors and company growth because it will allow information on a company's performance to be more detailed and detailed. We c also expand sample types and use samples of types of companies that have more diverse characteristics than manufacturing companies.

REFERENCES

Normayanti. (2017) "Pengaruh Kebijakan Hutang, Kebijakan Dividen Dan Profitabilitas Terhadap Nilai Perusahaan (Studi Empiris Pada Perusahaan Food and Beverage Yang Terdaftar Di Bursa Efek Indonesia)." *eJournal Administrasi Bisnis*, 5(2).

Pantow, M. S. R., Murni, S., and Trang, I. (2015) "Analisa Pertumbuhan Penjualan, Ukuran Perusahaan, Return On Asset, dan Struktur Modal Terhadap Nilai Perusahaan yang Tercatat di Indeks LQ 45." *Jurnal EMBA: Jurnal Riset Ekonomi, Manajemen, Bisnis dan Akuntansi*, 3(1).

Sasti, I., Rina Tjandrakirana, R., and Ilham, I. (2015) "Pengaruh Kebijakan Dividen, Kebijakan Hutang, Dan Profitabilitas Terhadap Nilai Perusahaan." *Jurnal Akuntanika*, 1(2).

Widiastari, P. A., and Yasa, G. W. (2018) "Pengaruh Profitabilitas, Free Cash Flow, dan Ukuran Perusahaan Pada Nilai Perusahaan." *E-Jurnal Akuntansi*, pp. 957–981.

Facing Global Digital Revolution – Nirmala Arum Janie, Dwi Mulyaningsih & Wahyu Rachmawati (eds)
 ISBN 978-0-367-33912-8

The determinant model of student entrepreneurship interest in Semarang University students

Nunik Kusnilawati Aprih Santoso, Nurhidayati & Ratna Wijayanti
Universitas Semarang, Kota Semarang, Indonesia

ABSTRACT: The study aims to analyze and empirically test the effects of the following: (1) attitudes, subjective norms, and perceived controls toward entrepreneurship education; (2) attitudes, subjective norms, and perceived controls toward entrepreneurial interests; and (3) entrepreneurship education toward interest in entrepreneurship at the University of Semarang. The sampling technique uses purposive proportional sampling. Data collection is done through a questionnaire. The analysis technique used is the SEM analysis method (Structural Equation Model) where PLS 3.0 software is used. The results of the study show that: (1) Attitudes, subjective norms, perceived controls influence entrepreneurship education. (2) Attitudes, subjective norms, perceived controls influence entrepreneurship interest. (3) Entrepreneurship education affects the interest in entrepreneurship. Thus the hypothesis is accepted.

Keywords: attitudes, subjective norms, perceived control, entrepreneurship education, entrepreneurial interest

1 INTRODUCTION

Population growth and change in the face of industry 4.0 have added new problems in the field of employment. At present, unemployment is a crucial concern being faced in labor issues in Indonesia. The demographic bonus that will be experienced by Indonesia starting in 2020 increasingly demands readiness to solve the problem of channeling a productive workforce, including those who have a higher education background.

Unemployment occurs because of an imbalance between the number of job seekers and available jobs: it occurs not only in Indonesia, but also throughout the world, and in various sectors including industry, mining, transportation, and others (Saiman, 2009). Data from the Central Statistics Agency (BPS) in August 2015 showed that the total open unemployment rate was 7.56 million, while February 2016 data shows unemployment of college graduates increased from 5.34 percent to 6.22 percent. Entrepreneurship is one alternative solution that can be used to solve the problem of unemployment, including of those who are highly educated. The number of entrepreneurs in Indonesia is only 1.56% of the total population, and this is relatively too small. Mc Clelland states, a country can prosper if there are at least 2% entrepreneurs. Japan has 2% entrepreneurs at the middle level and 20% at the small business level. Malaysia has 5% entrepreneurs, Singapore has 7% entrepreneurs, and America has more than 12% of the population become entrepreneurs.

The low number of entrepreneurs in Indonesia is considered related to the problem of interest in entrepreneurship. Several factors are assumed to influence the relatively low interest in entrepreneurship in the majority of the workforce in Indonesia, including those who are college graduates.

2 LITERATURE REVIEW

Hurlock (2004) argues that interest is a source of motivation that encourages people to do something they want if they are free to choose. When they see that something is profitable, they feel interested. This then brings satisfaction. When satisfaction decreases, interest decreases. According to H. C. Witherington quoted by Suharsimi (2009), interest is a person's awareness of an object, problem, or situation that has a connection to him. This limitation further clarifies the notion of interest in relation to one's attention. The emergence of more interest is due to the presence of a stimulant. According to Crow and Crow (1984), there are three factors that generate interest, namely, factors that arise from within the individual, social motive factors, and emotional factors. Interest in entrepreneurship college graduates is assumed to be influenced by (1) attitude, (2) subjective norms, (3) perceived control, and (4) entrepreneurship education.

Entrepreneurship education begins with the formation of an entrepreneurial mindset followed by the formation of creative and innovative behaviors in order to be creative (Morris, Lewis, and Sexton, 1994). Entrepreneurship education in higher education is related to building entrepreneurial character, entrepreneurial mindset, and entrepreneurial behavior that is always creative and innovative, creating added value or values, taking advantage of opportunities, and taking risks. Attitudes are interpreted as a reaction or response that arises from an individual to an object which then raises individual behavior toward the object in certain ways. Attitude is the perspective of someone who is positive, negative, or ambiguous toward a condition or condition that can affect the response these individuals (Ajzen et al., 2005). Attitudes can also be learned or developed from someone's experience with the environment. Subjective norms are individual beliefs about the demands and desires of people who are considered important. These demands are in the form of behavior that should be done and not done by individuals. People who are considered important can be parents, friends, colleagues, teachers, leaders, and others. Subjective norms can influence the intention of entrepreneurship from the external side in the form of support for the environment, both family and campus environment.

According to Icek (Ajzen, 2013) perceived control (perceived behavioral control) affects intention. This is based on the assumption that controls perceived by individuals will have implications in the form of motivation toward the person. The greater the perceived control of behavior by someone, the greater the intention to bring up a certain behavior.

3 METHODS, DATA, AND ANALYSIS

3.1 *Research methods*

The population in this study was students who had taken entrepreneurship courses from six faculties at the University of Semarang, totaling 22735 students. The data processing uses SEM (Structural Equation Model), using PLS 3.0 software, so the determination of the sample refers to Current (2014), which states that SEM analysis requires a sample of at least five times the number of indicators used. Based on the reference, the sample used is 5 X 21 = 105, which is distributed based on the comparison of the number of students per faculty, so that the sample is law faculty students (10.48%), economics faculty students (41.9%), engineering faculty students (18.1%), agricultural technology faculty students (4.76%), psychology faculty students (7.62%), and information and communication technology faculty students (17.14%).

The research instrument used was a questionnaire with two types of questions (closed and open). Dimensions of interest measurement include enthusiasm, experimentation, seeking opportunities, entrepreneurial obsession, and independence orientation. The measurement dimensions of entrepreneurship education include entrepreneurial information, entrepreneurial experience, entrepreneurial training, and community membership. The dimensions of attitude include cognitive, affective, psychomotor, and stimulus objects. The dimensions of subjective norms include family, social, environmental, and government. The perceived dimensions of control include perceptions of one's own strengths, weaknesses, opportunities, and threats.

4 RESULTS AND DISCUSSION

Respondents consisted of 43% men and 57% women, aged between 17 and 26 years of age, most of whom (45.7%) were aged 21–22 years. The rest were, in descending order, 19–20 years old (36.2%), 23–24 years old (8.6%), 17–18 years old (7.6%), and 25–26 years (1.9%). While the study stages of the respondents were divided into semesters 5–6 (46.67%), semesters 7–8 (43.81%), semesters 3–4 (5.71%), and semesters 1–2 (3.81%).

Convergent validity analysis shows that each indicator is able to explain the variable under study. This is because the original sample estimate value is high (above 0.5). The standard deviation obtained is also small, which indicates that there is uniformity of respondents' answers. Results of data processing average variance extracted (AVE) produces AVE for attitude variables of 0.684; subjective norms of 0.721; perceived control of 0.630; entrepreneurship education of 0.706; and variable interest in entrepreneurship of 0.605. At the critical limit of 0.5, the indicator in each construct is valid with the other items in one measurement.

Constructions are declared reliable if the composite reliability value is above 0.70 and Cronbach's alpha is above 0.60. In this study, the Smart PLS output of all constructs has a composite reliability value above 0.70 and Cronbach's alpha above 0.60. So it can be concluded that the construct has good reliability. From the calculation of R square, the value of $R2 = 1 - (1 - 0.878)(1 - 0.873) = 0.984$ means that the model is able to explain the phenomenon of entrepreneurial interest by 98.4%, while the remaining 1.6% is explained by other variables.

Inner model test or structural model test is used to evaluate relationships between latent constructs such as those that have been hypothesized in the study. Before testing the hypothesis, it is known that the t-table value for the confidence level of 95% (α of 5%) and the degree of freedom (df) = n-2 = 105–4 = 101 is equal to 1.66. Then the results of hypothesis testing get the following results:

Attitude influences entrepreneurship education: based on the results of data processing, the results are 0.321 with numbers t calculated at 6.153 where t count is greater than t table, which means the hypothesis is accepted.

Attitude influences interest in entrepreneurship: the parameter for attitudes toward entrepreneurship interest is 0.276, with t count 4.664, which means the hypothesis is accepted.

Subjective norms influence entrepreneurship education: parameters for subjective norm variables affecting entrepreneurship education are 0.409 and t count 8.337, this means that subjective norms influence entrepreneurship education. Thus the hypothesis is accepted.

Subjective norms influence the interest in entrepreneurship: parameters for subjective norms affect the interest in entrepreneurship by 0.195 with t count of 3.007, which is greater than t table, and thus the hypothesis is accepted.

Control perceived is influential on entrepreneurship education: control variables are perceived to have an effect on entrepreneurship education at 0.302 and t count 5.543 greater than t table, so the hypothesis is accepted.

Perceived control influences interest in entrepreneurship: estimated parameter for testing the perceived influence of control on the interest in entrepreneurship with a path coefficient of 0.155, can be accepted because the value of statistics = 2.466 is greater than the value of t table of 1.66, so the hypothesis is accepted.

Entrepreneurship education influences the interest in entrepreneurship: data processing results are 0.387 and t count is 3.565. This means the hypothesis is accepted.

5 CONCLUSION AND RECOMMENDATION

5.1 *Conclusion*

Attitude influences entrepreneurship education, and also interest in entrepreneurship. Subjective norms influence entrepreneurship education, as well as interest in entrepreneurship. Pperceived control influences entrepreneurship education, and also interest in entrepreneurship. Entrepreneurship education affects interest in entrepreneurship. Thus the overall hypothesis is accepted.

5.2 *Recommendation*

There needs to be variation in using fun, creative, and innovative learning methods, providing facilities and infrastructure to support entrepreneurship activities, and utilizing entrepreneurship programs such as entrepreneurial student programs (PMW) and student creativity programs (PKM). In addition to improving the quality of work practices in a study company through the sincerity of students in carrying out practices.

REFERENCES

Ajzen, I. et al. (2005) *The Handbook of Attitudes*. Mahwah, NJ, US: Lawrence Erlbaum Associates Publishers, pp. 173–221.

Ajzen, I. (2013) *Organizational Behavior and Human Decision Processes: The Theory of Planned Behavior*.

Crow, L. D., and Crow, A. (1984) "Psikologi pendidikan." *Buku l, Cetakan l, Surabaya: Bina Ilmu.*

Hurlock, E. B. (2004) "Psikologi perkembangan: suatu perkembangan sepanjang rentang kehidupan." *Edisi Keenam, Jakarta: Penerbit Erlangga.*

Morris, M. H., Lewis, P. S. and Sexton, D. L. (1994) "Reconceptualizing entrepreneurship: an input-output perspective." *SAM Advanced Management Journal*. Society for the Advancement of Management, 59(1), p. 21.

Suharsimi, A. (2009) "Dasar-dasar evaluasi pendidikan." *Jakarta: Bumi Aksara.*

Facing Global Digital Revolution – Nirmala Arum Janie, Dwi Mulyaningsih & Wahyu Rachmawati (eds)

Competitive advantage: The effect on the performance of MSME

Christian Suprapto, Paulus Wardoyo & Endang Rusdianti
Universitas Semarang, Kota Semarang, Indonesia

ABSTRACT: MSMEs (Micro, Small, Medium Enterprises) have a strategic position in the Indonesian economy, but they have problems in terms of capital and business management capabilities. The purpose of this research is to build a model that is useful for improving performance. This research uses all MSMEs businessmen in the batik sector in Rembang Regency with 135 respondents. Completion of structural equations, using Lisrel version 8.8. All research hypotheses are accepted. This research can prove that MSME business people need to have insight into entrepreneurial orientation and customer orientation. Because both are useful for companies, in building competitive advantage. However, customer orientation and competitive advantage play a big role in improving UKMK performance.

Keywords: Batik, customer orientation, Competitive Advantage and Performance

1 INTRODUCTION

The entrepreneurial orientation which is a classic problem of MSMEs is weak in terms of capital and managerial aspects (management, production, marketing, and human resource capabilities) and in the packaging of the product itself. Besides, at present, organizations that implement customer orientation are required to understand their customers' needs and their direction, change this understanding into actions in all functions in the organization, and respond to their understanding from customer needs. This is often a problem for customers who lack a basis for structured customer information.

Research into the influence of entrepreneurial orientation has been carried out, but still leaves controversy. Some say entrepreneurial orientation influences company performance (Mahmood and Hanafi. 2013, Jalilian et al. 2013, Zaini et. Al, 2014, and Wardoyo et. Al, 2015). Instead researchers such as Setyawati and Abrilia (2014) and Pardi et al (2014) state that entrepreneurial orientation does not affect company performance. Meanwhile, other researchers say customer orientation has an influence on company performance (Wulandari, 2012; Maurya.et.al, (2015), Fernández and Pinuer (2016), this is different from Rachmat's research (2006). Therefore, the problem of research is how to build a research model that can be used to improve the performance of MSMEs by using competitive advantage variables as a variable intervening.

Business performance is the result of a business organization's business and is seen from the results achieved. According to Mahmood and Hanafi (2013), MSME performance indicators are sales growth, profit growth, and asset growth. the growth in sales is an average growth of sales seen from the turnover of the company. the profit growth was an average growth of corporate profits. growth in assets was growth in average assets.

Competitive advantage is the advantages of the company for its competing rivals it is possible to generate more sales or margins or keep more customers than competitors (Salmones & Yin, 2014). Indicators of competitive advantage according to Fatah (2012) are product uniqueness, product quality, and competitive prices. The uniqueness of the product is the uniqueness possessed by the products produced by the company so that it distinguishes it from competing for products or general products on the market. As a breakthrough produced by the company in pouring the results of ideas or ideas to create something different or unique from others (competitors) so that they can have attractiveness for customers

Lumpkin and Dess (1996) Wiklund and Shepherd (2005) state that entrepreneurial orientation and organizational culture are closely related to the process of formulating strategies as the basis for business organization decision making. Whereas Miller and Fneseri (1982) revealed that entrepreneurial orientation can affect business performance. Entrepreneurial orientation refers to the processes, practices, and decision-making that encourage new inputs related to three aspects of entrepreneurship, namely taking risks, acting proactively and always being innovative (Lumpkin and Dess, 1996). Covin and Slevin (1991 stated that entrepreneurial orientation that is increasingly high can improve the ability of companies to market their products towards better business performance. Innovative, proactive, and risk-taking companies tend to be able to improve business performance. Besides, entrepreneurial orientation can also affect organizational strategy (Wardoyo et al. 2015). Daring to take risks is an entrepreneur's attitude that involves the willingness to tie up resources, dare to face the challenges of exploiting business strategies with the possibility of uncertain outcomes (Keh et al 2002 in Wardoyo et al, 2015). Proactivity reflects the desire of entrepreneurs to dominate competitors through combinations and aggressive and proactive movements, such as introducing new products or services over competitors to anticipate future demand and create environmental changes. Innovative refers to the attitude of entrepreneurs to be involved creatively in the process of experimenting new ideas that allow generating new production methods to produce new products or services, both for the current market and for new markets. A high entrepreneurial orientation is the main driver of profit so entrepreneurs have the opportunity to take advantage of the emergence of opportunities which in turn has a positive effect on business performance (Wiklund in Wardoyo et al, 2015).

Customer observation is the acquisition and use of information about events, trends, and relationships of customers in an organization, which is the knowledge that will be able to assist management in planning future actions. (Choo in Zaini et al, 2014). The company's control of customers and the company's ability to meet customer needs will make the company a step ahead of its competitors to increase competitiveness (Fatah, 2012). This description is consistent with Setiawan's (2015) research, that customer orientation has a positive effect on competitive advantage.

Wulandari Research (2012) which examines the influence of customer orientation on company performance with results that indicate that customer orientation influences the performance of MSMEs. The more successful a company adapts to changes in surrounding customers, the smaller the difference between the products produced by the company and the needs of consumers so that consumer growth will occur which is an indicator of company performance.

Maurya et al (2015) conducted a study on the effect of customer orientation on MSME performance, the result was that customer orientation had a positive effect on company performance.

Mahmood and Hanafi (2013), examining entrepreneurial orientation and business performance in small and medium-sized businesses owned by women in Malaysia also stated that competitive advantage has a positive effect on MSME performance. The higher the company's competitive advantage, the company will be superior to other companies that have similar products and will increase sales and company performance.

The results of the Pardi et al (2014) study are that competitive advantage has a positive effect on the performance of MSMEs. Increasing competitive advantage by companies is a process of creating a product that is not owned by another company. This makes consumers will be more interested in making purchases to improve their performance. Zaini et al (2014) which examined the effect of entrepreneurial orientation on performance with a competitive advantage as a mediator had results that stated that competitive advantage had a positive effect on the performance of MSMEs.

2 RESEARCH METHODS

The population of this study was the MSMEs unit of Batik Tulis in Rembang Regency, which amounted to 135 entrepreneurs. This study uses the census method. Data collection carried out by using questionnaires distributed to respondents. The scale of measurement is done using a scale of one for answers that strongly disagree, up to seven for answers strongly agree. Completion of structural equations in research using Lisrel software version 8.8

3 RESULTS AND DISCUSSION

Validity testing is a test that aims to determine the ability of an indicator in measuring the latent variable (Ghozali and Fuad, 2008). To test the validity of each indicator, Confirmatory Factor Analysis (CFA) is used. The calculation results show that all indicators have a value greater than t table (1.65639). All indicators are valid, the reliability test is carried out using composite reliability

Reliability test results show all variables have ρ (composite reliability) greater than 0.9 which means reliable. The results of data processing are as follows:

Structural Equations

KB = 0.13 * OK + 0.86 * OP, Errorvar. = 0.066, Rý = 0.93

(0.058)	(0.077)	(0.013)
2.21	11.15	5.03

KIN = 0.65 * KB + 0.15 * OK + 0.22 * OP, Errorvar. = 0.015, Rý = 0.98

(0.095)	(0.040)	(0.092)	(0.0061)
6.89	3.60	2.36	2.50

Note:
KB = Competitive Advantage, OK = Entrepreneurship Orientation, OP = Customer Orientation and KIN = Performance

Image results of data processing appear as follows:

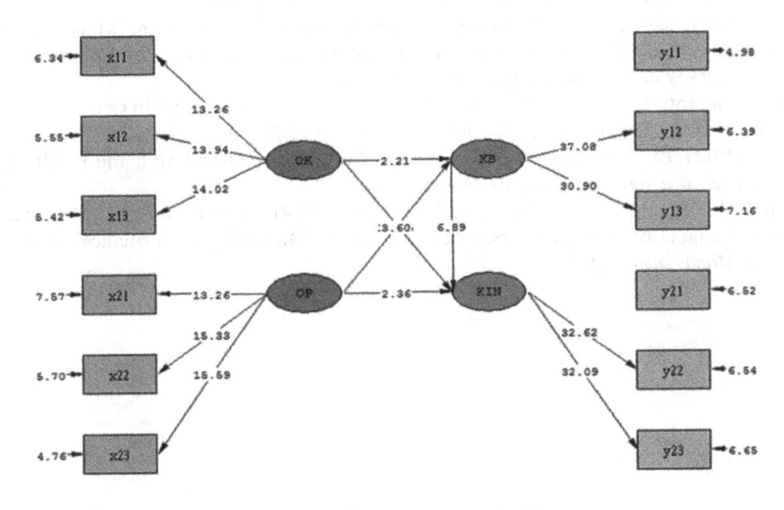

Figure 1. Full model.

The research hypothesis is accepted because the value of t count is greater than t table. The calculation results show that the total effect of customer orientation and competitive advantage is greater than the total effect of entrepreneurial orientation and competitive advantage

4 CONCLUSION

This research successfully proved that, in improving the performance of MSMEs, entrepreneurs are not only entrepreneurial oriented. However, businesses must be more customer-oriented. By orienting to customers, businesses will become more sensitive to what customers need, be able to quickly deal with customer complaints, be friendly and polite. Besides, business actors can also evaluate the services and products produced, and in the future improve/ improve them. What is done, all aims to produce competitive advantages that are realized in the form of unique, quality products and can become an icon for the local. The low level of education of business actors is an obstacle to data collection, especially in digesting research instruments.

REFERENCES

Fernandez, L. V., & Pinuer, F. V. (2016) 'Influence of Customer Orientation, Brand Value and Business Ethics Level on Organizational Performance', *Review of Business Management*, 18(59).

Jalilian, K., Navid, B. J., & Ghanbary, M. (2014) 'The Impact of Entrepreneurial Orientation and Market Orientation on the Performance of Industrial Firms Listed in Tehran Stock Exchange (Food, Chemical, Pharmaceutical, Automobile)', *International Research Journal of Applied and Basic Sciences*, 5(5).

Mahmood, R., & Hanafi, N. (2013) 'Entrepreneurial Orientation and Business Performance of Women-Owned Small and Medium Entreprises in Malaysia: Competitive Advantage as Merdiator', *International Juornal of Business and Social Science*, 4(1).

Maurya, U., Mishra, P., Anand, S., & Kumar, N. (2015) 'Corporate identitiy, Customer Orientation and Performance of SMEs: Exploring the Linkage', *IIMB Management Review*, 20(1).

Pardi, S., Imam, S., & Zainul, A. (2014) 'The Effect of Market Orientation and Entrepreneurial Orientation toward Learning Orientation, Innovation, Competitive Advantages and Marketing Performance', *European Journal of Business and Management*, 6(21).

Rachmat, B. (2006) 'Customer Orientation, Market Orientation and Innovation and Its Influence on Business Performance of Three-Star Hotels in Indonesia Equity', 11(3).

Setyawati, Abrilia H. (2014) 'The Influence of Entrepreneurship Orientation and Market Orientation on Company Performance Through Competitive Advantage and Perception of Environmental Uncertainty (Trade Survey in Kebumen Regency)', *JAAI*, 10(1).

Wardoyo, P., Rusdianti, E., & Purwantini, S. (2015) 'Effect of Entrepreneurship Orientation on Business Strategy and Business Performance', *Journal & Proceeding FEB Unsoed*, 5(1).

Wulandari, A. (2012) 'Effect of Customer Orientation, Competitive Orientation and Product Innovation on Performance', *Management Analysis Journal*, 1(2).

Zaini, A., Hadiwidjojo, D., Fatchurohman, & Maskie, G. (2014) 'Effecto of Competitive Advantage As A Mediator Variabel of Entrepreneurship Orientation to Marketing Performance', *IOSR Journal of Business and Management*, 16(5).

Facing Global Digital Revolution – Nirmala Arum Janie,
Dwi Mulyaningsih & Wahyu Rachmawati (eds)
 ISBN 978-0-367-33912-8

Taxpayer considerations when utilizing Indonesia's tax amnesty policy

N. Herawati
Student in the Doctoral Program of Economics Science, University of Sebelas Maret, Kota Surakarta, Indonesia
Faculty of Economics and Business, University of Trunojoyo Madura, Bangkalan, Indonesia

Rahmawati Bandi & D. Setiawan
Faculty of Economics and Business, University of Sebelas Maret, Kota Surakarta, Indonesia

D. Indudewi
Universitas Semarang, Kota Semarang, Indonesia
Student in the Doctoral Program of Economics Science, University of Sebelas Maret, Kota Surakarta, Indonesia

ABSTRACT: This study examined the factors that taxpayers consider when utilizing Indonesia's tax amnesty policy. The research sample consisted of 272 taxpayers in Surakarta, Semarang, and Pamekasan. Sampling was done incidentally, and the analytical method used was factor analysis. The results revealed that taxpayers consider seven factors – i.e., consequences for taxpayers, the benefits of tax amnesty, regulatory certainty, taxpayers' confidence, rewards for tax compliance, time, and sanctions.

Keywords: tax amnesty, taxpayer, Indonesia, factor analysis

1 INTRODUCTION

The Indonesian government implemented a tax amnesty policy in 1984. However, the policy was considered not very successful given the response from taxpayers and the modernization of the taxation system in Indonesia (Haryanto, 2016; Laksana, 2005). The tax amnesty policy, now named the sunset policy, was again carried out in 2008. This sunset policy provided for eliminating administrative sanctions for taxpayers who are underpaid and make mistakes in filling out their tax returns. The year 2015 was designated as Taxpayer Development Year, for the elimination of taxation and administrative sanctions for taxpayers who have not adequately complied with tax regulations (Direktorat Jenderal Pajak, 2015). In 2016, the government once again re-implemented the tax amnesty policy.

Under the 2016 tax amnesty policy, the government targeted the ransom of the tax amnesty program of 165 trillion rupiah. The tax amnesty policy was divided into three periods: period I (from the date of promulgation as of September 30, 2016), period II (from October 1, 2016, until December 31, 2016), and period III (from January 1, 2017, until March 31, 2017). The total ransom of period I reached 97.1 trillion rupiah (Adityowati, 2016). This means that the tax amnesty policy in period I did not reach the expected target, even though it provided several benefits for taxpayers, including the elimination of owed taxes, administration sanctions, and criminal sanctions on assets acquired in the year 2015 (Direktorat Jenderal Pajak, 2015).

It appears that the benefits the policy offered were not enough to encourage taxpayers to use them. Then what other factors did taxpayers consider when utilizing the tax amnesty policy? The existing research has examined the influence of the tax amnesty policy on taxpayer compliance (Agbonika, 2015; Siregar, 2011; Sudarma, 2017), what happens after tax amnesty (Okfitasari, Meikhati, and Setyaningsih, 2017), tax amnesty and its implementation in several countries (Santoso and Setiawan, 2009), the implementation of tax amnesty in Indonesia (Dewi, 2016), and tax amnesty in Indonesia from an accounting perspective (Natania and Davianti, 2018). Therefore, this study aimed to explore the factors that taxpayers consider when utilizing the tax amnesty policy. Knowing what factors taxpayers consider will provide support to policy makers – namely, the directorate general of taxes – in making further policies. Thus, this information is expected to increase taxpayers' participation in the amnesty tax policy and therefore to raise state revenues.

2 RESEARCH METHODS

2.1 *Population and samples*

The population of this study comprised taxpayers at Surakarta KPP Pratama, Pamekasan KPP Pratama, and several KPPs in Semarang. The researchers carried out incidental sampling. The details of the research sample selection were as follows.

Table 1. Research samples.

Number	Tax office	Number of respondents	Total samples
1.	Surakarta	200	
2.	Pamekasan	51	
3.	Semarang	32	283
4.	Incomplete Questionnaires	(11)	272

2.2 *Data collection*

Data collection was completed using a questionnaire. The questionnaire contained respondent information and 22 attributes/indicators based on several sources (attributes A–F (www.pajak.go.id/amnestipajak), attribute G (Ortax.org, 2016), attributes H and K (Law No. 11 of 2016, Article 18, paragraph [3]), attribute L (Law No. 11 of 2016, Article 5, paragraph [1]), attribute M (Law No. 11 of 2016, Article 5, paragraph [3]), attributes N and O (Haryanto, 2016), attributes Q–S (Akbar, 2016), and attributes T–V (Karo, 2016).

2.3 *Data analysis*

The analysis was carried out using factor analysis. The factor analysis employed in this study was exploratory factor analysis, which involves looking for a number of indicators in order to form general factors without any prior theoretical foundation. The steps in carrying out factor analysis followed Widarjono (2015): (1) Calculate the correlation matrix (the Kaiser-Meyer-Olkin measure of sampling adequacy (KMO MSA), or Bartlett's sphericity test). (2) Look for factors or factor extraction. (3) Look for factors that can optimize the correlation between the independent indicators observed.

3 RESULT AND DISCUSSION

3.1 *Respondents' characteristics*

The respondents' characteristics can be seen in Table 2.

Table 2. Respondents' characteristics.

Number	Characteristics	Amount	Number	Characteristics	Amount
1	Gender		3	Period of tax amnesty	
	Male	160		I	66
	Female	112		II	61
2	Taxpayer			III	145
	Individual taxpayer – micro, small, and medium enterprises (MSME)	100	4	Tax office area	
	Individual taxpayer – non-MSME	116		Surakarta	193
	Corporate taxpayer – MSME	28		Pamekasan	47
	Corporate taxpayer – non-MSME	28		Semarang	32

3.2 *Factor analysis results and discussion*

The calculation result of the correlation matrix (Table 3) showed that the KMO MSA is above 80%, which means that it meets the requirements for factor analysis. Bartlett's sphericity test showed statistically significant results, which means that the correlation matrix is not an identity matrix, so factor analysis can be used.

The next step was factor extraction. Factor extraction results revealed seven components/factors with an initial total eigenvalue ≥ 1; these seven factors can explain the indicators well, so they must be included. The first factor explains 29.132% of the variation, the second factor 10.745%, the third 7.761%, the fourth 6.590%, the fifth 5.747%, the sixth 4.677%, and the seventh 4.589%. All seven factors can explain 69.241% of the variation.

The first factor includes exemption from income tax related to the process of reversing the name of assets (F); impacts arising related to tax obligations after the tax amnesty policy (G); information on the tax amnesty policy (J); the ransom rate (K); the amount of assets that must be reported in the tax amnesty policy (M); and the amount of the tax amnesty policy (P). The second factor includes the elimination of owed taxes (A); termination of the inspection process, examination of preliminary evidence, and investigation (D); which can be used as a base for investigating other crimes (E); and taxpayers' understanding of the tax amnesty policy (I). The third factor includes the regulations related to tax amnesty (N); the baseline in implementing the tax amnesty (O); and how the policy is similar to tax amnesty issued in the future (Q). The fourth factor includes taxpayers' confidence in this program (U) and lack of knowledge of how to exploit it (V). The fifth factor includes tax administrative sanctions and tax criminal sanctions (B) and no inspection, the examination of preliminary evidence, and investigation (C). The sixth factor includes no appreciation for tax compliance (R) and lack of time (T). The seventh factor includes a 200% increase in the income tax rate if income taxes are not paid after the tax amnesty policy ends (H).

The results of the study are depicted in Table 4. We labeled seven factors formed from 22 attributes. Some of these variables have also been used by Dewi and Noviari (2017) and Wirawan and Noviari (2017), but some variables have not been used in previous studies. These factors can be a concern for policy makers in implementing the tax amnesty program going forward. The directorate general of taxes can provide detailed information about the tax amnesty program, pay attention to rewards for obedient taxpayers, and promote the ease of joining the program and the certainty of regulations and sanctions.

Table 3. KMO MSA and Bartlett's test.

Kaiser-Meyer-Olkin measure of sampling adequacy		0.818
Bartlett's test of sphericity	Sig.	0.000

Table 4. Factors that taxpayers consider when utilizing the tax amnesty policy.

Factor	Attribute	Name of the factor
1	(F), (G), (J), (K), (M), (P)	Consequences for taxpayers
2	(A), (D), (E), (I)	Understanding the benefits of tax amnesty
3	(N), (O), (Q)	Regulatory certainty
4	(U), (V)	Taxpayers' confidence
5	(B), (C)	Benefits of tax amnesty
6	(R), (T)	Reward for tax compliance and consideration of lack of time
7	(H)	Sanction

4 CONCLUSION

This study aimed to identify the factors that taxpayers consider when utilizing the tax amnesty policy, based on 22 attributes, and found that taxpayers largely take into account seven factors in following the 2016 tax amnesty program. This research was based in three regions. Future research can be developed by expanding the survey area.

REFERENCES

Adityowati, P. (2016). Tebusan amnesti pajak periode I tembus Rp 97,1 Triliun. https://m.tempo.co

Agbonika, J. (2015). Tax amnesty for delinquent taxpayers: A cliché in Nigeria. *Global Journal of Politics and Law Research*, 3(3), pp. 105–120.

Akbar, D. S. F. (2016). Tax amnesty dan momentum reformasi. www.kemenkeu.go.id/

Dewi, A. S. (2016). Analisis implementasi pengampunan pajak (tax amnesty) di Indonesia. *Publik*, 1(1), pp. 94–105.

Dewi, N. P. A. and Noviari, N. (2017). Pengaruh kesadaran wajib pajak, pelayanan fiskus, dan sanksi perpajakan pada kemauan mengikuti tax amnesty. *E-Jurnal Akuntansi*, 19(2), pp. 1378–1405.

Direktorat Jenderal Pajak (2015). Ini bedanya sunset policy 2008 vs TPWP 2015. www.pajak.go.id/

Haryanto, J. T. (2016). Tax amnesty dan kinerja perpajakan 2016. www.kemenkeu.go.id/

Karo, A. J. K. (2016). Survey: Tax amnesty berpotensi gagal !! www.creative-trader.com/

Laksana, H. Y. A. (2005). Belajar dari Afrika Selatan soal amnesti pajak. www.pajak2000.com/

Natania, E. S. and Davianti, A. (2018). An accounting perspective of tax amnesty in Indonesia. *Journal of Accounting Auditing and Business*, 1(1), pp. 1–18.

Okfitasari, A., Meikhati, E., and Setyaningsih, T. (2017). Ada apa setelah tax amnesty? *Jurnal Akuntansi Multiparadigma*, 8(3), pp. 511–527.

Ortax.org (2016). Sebelum mengajukan tax amnesty, pahami aspek pajak yang timbul. www.ortax.org/

Peraturan Menteri Keuangan Republik Indonesia Nomor 118/PMK.03/2016 Tentang Pelaksanaan Undang-Undang Nomor 11 Tahun 2016 Tentang Pengampunan Pajak.

Santoso, U. and Setiawan, J. M. (2009). Tax amnesty dan pelaksanaannya di beberapa negara: Perspektif bagi pebisnis Indonesia. *Sosiohumaniora*, 11(2), p. 111–125.

Siregar, B. (2011). Pengaruh layanan fiskus dan pelaksanaan sunset policy terhadap kepatuhan wajib pajak dalam upaya peningkatan pajak. *Jurnal Ekonomi & Bisnis*, 5(2), pp. 173–189.

Sudarma, I. (2017). Does voluntary tax compliance increase after granting tax amnesty? *GART Journals afr138, Global Academy of Training and Research (GATR) Enterprise*.

Undang-Undang Republik Indonesia Nomor 11 Tahun 2016 Tentang Pengampunan Pajak.

Widarjono, A. (2015). *Analisis Multivariat Terapan*. Yogyakarta: UPP STIM YKPN.

Wirawan, I.B.N.A.P. and Noviari, N. (2017) Pengaruh penerapan kebijakan tax amnesty dan sanksi perpajakan terhadap kepatuhan wajib pajak orang pribadi. *E-Jurnal Akuntansi*, 21(3), pp. 2165–2194.

Facing Global Digital Revolution – Nirmala Arum Janie, Dwi Mulyaningsih & Wahyu Rachmawati (eds)
 ISBN 978-0-367-33912-8

Social preferences in budget decision making

Dian Indudewi
Student of Doctoral Program of Economics Science, University of Sebelas Maret, Surakarta, Indonesia
Universitas Semarang, Semarang, Indonesia

Agung Nur Probohudono
Faculty of Economics and Business, University of Sebelas Maret, Surakarta, Indonesia

ABSTRACT: Local Government performance came into the spotlight when the Indonesian Corruption Watch (ICW) announced that the irrelevant use of funds became the highest factor of corruption throughout 2017. The loophole in the budget mechanism triggers the corruption. This research attempt to find the interconnection between budget actor social preferences and decision making in budget setting process. The research contributes both in principle and empirical. The results of this research indicate that budget actor social preferences have a significant effect on decision making in budget settings process. The findings also suggest the superior should give a concern to the conformity between activities and government plan, and how to improve subordinate sense of belonging that quite low to their institution.

Keywords: budget social preferences, budget setting, budget actor

1 INTRODUCTION

Mechanism of government budgeting in Indonesia has strict regulations. However, budget actors still have an opportunity to cheat in the budgeting process and even cause delays in budget approval (Febrina & Isril, 2016; Parwati, Budiasih, & Astika, 2015; Subechan, Hanafi, & Haryono, 2014; Riyanto, 2012; Hanida, 2010). The Supreme Audit Agency (BPK) found that out of 6,222 weaknesses in the internal control system in the regional government, 2,887 cases were caused by weaknesses in controlling the implementation of the revenue and expenditure budget. The results of BPK's examination are in line with the mapping carried out by ICW regarding cases of corruption in 2017. ICW uncovered that out of 576 corruption cases 154 cases were misused by the regional government.

A budget is a control device (Kelly, 2005) but the gap in the budget mechanism makes the budget an object that is vulnerable to manipulation (Amans, Mazars-Chapelon, & Villeseque-Dubus, 2015). The inability of local government revenues to finance all regional expenditures encourages budget actors to set priorities and select activity programs to be included in the budget. The process of selecting programs and activities is a process that is full of political interests. Febrina & Isril (2016) and Hanida (2010) see a difference in preferences between the executive and the legislature regarding budgeting due to differences in interests. Preferences can appear during the process of selecting programs and activities that will be included in the budget. Selection of preferences can be triggered by social relationships that are intertwined between organizational members and between organizations (Frederick, Loewenstein, & O'Donoghue, 2002).

Research related to government budgets in Indonesia so far referring to proving the opportunistic behavior of budget actors by looking at budget spreads (Parwati, et.al., 2015). Sujaie (2013) explored more in the opportunistic behavior of budget actors and found that there were bargaining and soft negotiations in the preparation of the East Java Provincial Budget. Research by Febrina & Isril, (2016) and Hanida (2010) found that there was a difference of interests between the executive and the legislature, which cause the late of budget settlement. Research outside Indonesia related to the budget also provides interesting results. Research on Costa-Font, Forns, & Sato (2015) revealed that social priorities and preferences are inherent in choosing programs and activities, individuals can change their views when they are involved in budgeting. The research of Ozdemir, Johnson & Whittington, (2016) and Mutiganda (2013) found that ideology plays a role in the selection of program and activity preferences, and funding sources are an important consideration when making policies.

The results of previous studies have indicated that social preferences can play a role during the process of selecting programs and activities. This study seeks to look at preferences that are the basis of the reasons for budget actors in choosing programs and activities and see the influence of social preferences on budgeting.

This research contributes both theoretically and practically. The theoretical contribution is a better understanding of the social preferences inherent in the individual influencing the decisions that will be made, while the practical contribution is to provide an overview of the social preferences of budget actors in choosing and determining programs and activities.

2 LITERATURE REVIEW AND THEORETICAL BACKGROUND

Social exchange derived from relationships and negotiations take place in the social order. The law of reciprocity (giving and receiving) forms the basis of all social relations (Gobel, Vogel, & Weber, 2013; Gouldner, 1960). As long as there is a social exchange within the organization, differences in interests cannot be avoided. To achieve organizational goals and bridge the differences between negotiations. Lawler (1992) stated that bargaining is a form of cooperative decision-making of diverse motivations and alternatives. Negotiations cannot be separated from social structure and power, where more minority parties try to minimize the negative impact of negotiations (Tangpong, C.,Li Jin., Ting Hung, 2016; Bottom, Holloway, Miller, Mislin, & Whitford, 2006; Cook & Emerson, 1978; Hamner, 1974). Social values become a counterweight that can create social balance (Cook, 2015; Cook & Rice, 2003; Cook, 1977; Emerson, 1976). These social values can be in the form of altruism (Zuckerman, Chen, & Nau, 2018; Murphy, Ackerman, & Handgraaf, 2011; Cropanzano & Mitchell, 2005), trust (Blay, Douthit, & Fulmer, 2018; Zuckerman, et.al, 2018), and fairness (Falk, Becker, Dohmen, Huffman, & Sunde, 2016; Cropanzano & Mitchell, 2005; Andreoni, Brown, & Vesterlund, 2002). Preference arises from within an individual as a form of addressing social exchanges that occur. Preference refers to the main choices made by an individual through considerations which basically tend to contain fulfillment of individual needs. Therefore, when individuals are faced with social choices that are around them, then the individual has motivational considerations that ultimately drop the choices that provide incentives for him (Jansson & Bursell, 2018; Falk, et.al. 2016; Andreoni et.al. 2002). Preferences can appear during the process of selecting programs and activities that will be included in the budget. The existence of priority setting when the budgeting process can trigger budget actors to choose (Ozdemir, 2016; Costa-Font, et.al. 2015; and Mutiganda, 2013).

3 METHOD

The research was carried out in the Central Java Provincial Government with two strong reasons: (1) the inaccuracy of realization of the 2017 budget by the BPK was discovered, and (2) a decrease in the revenue target of 22% in 2016 as a result of the previous year's high earnings estimates. Data is obtained by distributing questionnaires to budget compilers, namely the head of the work unit, planner, and program staff whose task is to compile the budget. The questionnaire is the development of several previous studies that examined the same topic qualitatively. The questionnaire was conducted in addition to the support of literature by conducting interviews with budget compilers. The informant is from 5 agencies, namely the regional development and planning agency, the regional asset and financial management agency, the inspectorate, the cooperative and small-medium enterprises services, and the industrial service. The questionnaire was then discussed again and also consulted with the grammar expert to minimize the understanding bias.

After the questionnaire was arranged, a pre-test was conducted to test the validity and reliability. There is eight indicators for budget social preferences such as: support and cooperation for all team compiler, negotiation and commitment, good name, trust, fairness and honesty, income feedback with strongly don't agree-strongly agree five Likert scale. Budget performance measured by indicators which confirm about target vs realization for: planning, input, output, absorption level, reward, and good name. Questionnaires that have passed the pre-test are then distributed to regional work unit in the Central Java provincial government. Those who can fill out the questionnaire are echelon 3 and 4 officials who are directly involved in planning and drafting the work unit budget. Testing to answer the hypothesis by using the partial least square method.

4 RESULT AND DISCUSSION

Based on 120 questionnaires distributed, 100 questionnaires have returned and there are 92 questionnaires that can be processed, 8 questionnaires cannot be used due to incomplete data. After tabulating the data, the instrument test is carried out, namely the validity and reliability test. Test the validity is valid if the value of the loading factor of at least 0.7, 0.5-0.6 allowed if the questionnaire is still a level of development. To measure reliability, composite reliability is used with values above 0.7. The results of the instrument test show that there are several indicators that do not pass the validity test so they must be excluded from the model, while for the reliability test results all variables show a number above 0.7.

After passing the instrument test then the next is a hypothesis test and obtained at a value of 12,276 with a significance value of 0,00 so that it can be concluded that the social preference budget has a significant effect on the budget setting.

5 CONCLUSION

Based on the test results it is known that social preference is a factor that influences the budgeting process. There are several other interesting findings to look at. First, the budget setting indicator "all programs and activities included in the budget in accordance with the regional government work plan" receive the lowest validity value. This is an interesting issue where programs and activities that are included in the budget should be in accordance with the work plan of the regional government. Second, budget actor's sense of belonging is quite low, judging from the low score of respondents' answers related to the indicator item "organization's good name". Both of these things should be of concern to the local government and can be a signal of what the Central Java provincial government has to do with the BPK findings. This study also has limitations, where there are several indicator values that are still in the range of 0.5-0.6 so that it is necessary to improve or develop construct measurement indicators. The time for distributing questionnaires needs to be considered so that the response rate becomes better.

REFERENCES

Andreoni, J., Brown, P.M. and Vesterlund, L. (2002) 'What makes an allocation fair? Some experimental evidence', *Games and Economic Behavior*, 40(1), pp.1–24.

Amans, P., Mazars-Chapelon, A. and Villesèque-Dubus, F. (2015) 'Budgeting in institutional complexity: The case of performing arts organizations', *Management Accounting Research*, 27, pp.47–66.

Blay, A., Douthit, J. and Fulmer III, B. (2019) 'Why don't people lie? Negative affect intensity and preferences for honesty in budgetary reporting', *Management Accounting Research*, 42, pp.56–65.
Bottom, W.P., Holloway, J., Miller, G.J., Mislin, A. and Whitford, A. (2006) 'Building a pathway to cooperation: Negotiation and social exchange between principal and agent', *Administrative Science Quarterly*, 51(1), pp.29–58.
Cook, K.S. (1977) 'Exchange and power in networks of interorganizational relations', *The sociological quarterly*, 18(1), pp.62–82.
Cook, K. S. (2015) 'Exchange: Social', *International Encyclopedia of the Social & Behavioral Sciences: Second Edition*, 8. Elsevier.
Cook, K.S. and Emerson, R.M. (1978) 'Power, equity and commitment in exchange networks', *American sociological review*, pp.721–739.
Cook, K.S., Cheshire, C., Rice, E.R. and Nakagawa, S. (2013) 'Social exchange theory, Handbook of social psychology', *Springer, Dordrecht*, pp. 61–88.
Costa-Font, J., Forns, J.R. and Sato, A. (2015) 'Participatory health system priority setting: Evidence from a budget experiment', *Social Science & Medicine*, *146*, pp.182–190.
Cropanzano, R. and Mitchell, M.S. (2005) 'Social exchange theory: An interdisciplinary review', *Journal of management*, 31(6), pp.874–900.
Emerson, R.M. (1976) 'Social exchange theory', *Annual review of sociology*, 2(1), pp.335–362.
Falk, A., Becker, A., Dohmen, T.J., Huffman, D. and Sunde, U. (2016) 'The preference survey module: A validated instrument for measuring risk, time, and social preferences.
Febrina, R., & Isril. (2016) 'Analisis Proses Politik Pembahasan dan Penetapan Anggaran Pendapatan dan Belanja Daerah (APBD) Provinsi Riau Tahun Anggaran 2016.
Frederick, S., Loewenstein, G. and O'donoghue, T. (2002) 'Time discounting and time preference: A critical review', *Journal of economic literature*, 40(2), pp.351–401.
Göbel, M., Vogel, R. and Weber, C. (2013) 'Management research on reciprocity: A review of the literature', *Business Research*, 6(1), pp.34–53.
Gouldner, A.W. (1960) 'The norm of reciprocity: A preliminary statement', *American sociological review*, pp.161–178.
Hamner, W.C. (1974) 'Effects of bargaining strategy and pressure to reach agreement in a stalemated negotiation', *Journal of Personality and Social Psychology*, 30(4), p.458.
Hanida, R.P. (2016) 'Dinamika Penyusunan Anggaran Daerah: Kasus Proses Penetapan Program dan Alokasi Anggaran Belanja Daerah di Kabupaten Sleman', *Jurnal Penelitian Politik*, 7(1), p.15.
Jansson, F. and Bursell, M. (2018) 'Social consensus influences ethnic diversity preferences', *Social influence*, 13(4), pp.192–208.
Kelly, J.M. (2005) 'A century of public budgeting reform: The "Key" question', *Administration & Society*, 37(1), pp.89–109.
Lawler, E.J. (1992) 'Power processes in bargaining', *Sociological Quarterly*, 33(1), pp.17–34.
Murphy, R.O., Ackermann, K.A. and Handgraaf, M. (2011) 'Measuring social value orientation', *Judgment and Decision making*, 6(8), pp.771–781.
Mutiganda, J.C. (2013) 'Budgetary governance and accountability in public sector organisations: An institutional and critical realism approach', *Critical Perspectives on Accounting*, 24(7-8), pp.518–531.
Ozdemir, S., Johnson, F.R. and Whittington, D. (2016) 'Ideology, public goods and welfare valuation: An experiment on allocating government budgets', *Journal of choice modelling*, 20, pp.61–72.
Parwati, S.M., Budiasih, I.G.A.N. and Astika, I.B.P. (2015) 'Perilaku Oportunistik Penyusun Anggaran', *Jurnal Ilmiah Akuntansi dan Bisnis*.
Riyanto, A. (2012) 'Politik Anggaran Provinsi Jawa Tengah: Analisis Realisasi APBD Provinsi Jawa Tengah Tahun Anggaran 2008-2010', *SPEKTRUM*, *12*(2).
Subechan, S., Hanafi, I. and Haryono, B.S. (2014) 'Analisis Faktor-faktor Penyebab Keterlambatan Penetapan APBD Kabupaten Kudus', *WACANA, Jurnal Sosial dan Humaniora*, 17(1), pp.35–46.
Sujaie, A.F. (2013) '*Oportunisme Perumus Kebijakan Anggaran Dalam Penyusunan Apbd Provinsi Jawa Timur Tahun 2013: Fenomena Dalam Pelaksanaan Belanja Hibah*', (Doctoral dissertation, Universitas Gadjah Mada).
Tangpong, C., Li, J. and Hung, K.T. (2016) 'Dark side of reciprocity norm: Ethical compromise in business exchanges', *Industrial Marketing Management*, 55, pp.83–96.
Zuckerman, I., Cheng, K.L. and Nau, D.S. (2018) 'Modeling agent's preferences by its designer's social value orientation', *Journal of Experimental & Theoretical Artificial Intelligence*, 30(2), pp.257–277.

Facing Global Digital Revolution – Nirmala Arum Janie, Dwi Mulyaningsih & Wahyu Rachmawati (eds)
© 2020 Taylor & Francis Group, London, ISBN 978-0-367-33912-8

University-led intellectual property commercialization: Cases of Malaysian universities

S.M. Sharif
Universiti Teknikal Malaysia Melaka (UTeM), Melaka, Malaysia

A. Isa
Universiti Teknologi MARA (UiTM), Kuala Lumpur Malaysia

A.Y.M. Noor
Universiti Kebangsaan Malaysia (UKM), Bangi, Malaysia

A.Z. Samsudin
Universiti Teknologi MARA (UiTM), Kuala Lumpur Malaysia

M.A.M. Nizah
Universiti Sains Islam Malaysia (USIM), Nilai, Malaysia

M.S.A. Azzis
Universiti Malaya (UM), Kuala Lumpur, Malaysia

ABSTRACT: Universities are at the center of innovation. The process through which innovation comes from universities to industries, also called intellectual property (IP) commercialization, tends to be affected by various factors and contexts. The broad objective of this study was to explore the current status of IP commercialization in Malaysian universities. This study aimed to examine IP commercialization's intensity, its processes, and the factors affecting it in selected Malaysian universities. University Sains Malaysia (USM) and Universiti Teknikal Malaysia Melaka (UTeM) were selected for convenience, and top-, middle-, and low-level university officials were interviewed using a semi-structured, in-depth interview protocol. The findings indicated that the intensity of IP commercialization depends on experience, industry linkage, properly planned research areas, and active researchers from various research institutes. This research will contribute to the existing understanding of IP commercialization in the Malaysian context, along with the necessary measures to take at the policy, industry, and university levels, in order to foster innovation and socioeconomic development in the country.

Keywords: intellectual property, commercialization, university, innovation

1 INTRODUCTION

Intellectual property (IP) is society's recognition of intellectual efforts. Commercialization of intellectual property is one of the challenging tasks for effective innovation management at universities (Sharif, Ahamat, et al., 2018). Malaysian public and technology-based universities still do not have a sufficient framework to promote timely commercialization of university-led innovations (Hamzah, 2006). Critical and strategic information and its constant regeneration in the new knowledge economy make it possible for corporations to seek targeted advantages (Alias et al., 2018). Ironically, its essential role is acknowledged only when an organization is mismanaged, or worse, collapses (Mat Isa, 2009).

The initiatives taken and models applied are all for the ease of IP commercialization, so that the value of innovation comes to users and creativity does not die (Wong, Ho, et al., 2007). People should take action and share their current understanding with others (Mat Isa et al., 2016). Overall, this study sought to highlight how we create innovations in universities and how we market them.

2 INTELLECTUAL PROPERTY COMMERCIALIZATION

As described in Article 2 of the World Intellectual Property Organization (WIPO) Convention (1967), the term *intellectual property* means "all rights resulting from intellectual activity in the industrial, scientific, literary or artistic fields. Intellectual property commercialization involves various steps or activities which may either be transactional or developmental concerning IP (Sharif, Nizam, et al., 2018).

2.1 *Steps in IP commercialization*

The commercialization of the IP process is concerned with five essential elements (Jaiya, 2008). These five elements consist of creation, protection, management formation, exploitation, and due diligence or evaluation. Protection is vital to prevent the loss of the IP or its value. Infringement may occur in the form of unauthorized disclosure, use, distribution, or exploitation. Ideally, IP should be protected from infringement from the moment it is created (Hamzah, 2006). Management of IP involves various steps such as identifying the IP, collating and recording the IP, and designing and implementing a database to manage all available IP (Whittaker, 2003). Exploitation means to utilize the IP in such a manner as to bring revenue to the enterprise. There are various methods of exploitation such as licensing, assignment, capitalization, collateralization, and securitization (Woodward, 2004).

2.2 *University intellectual property*

Universities are the sources of innovation, and extant works of literature on IP commercialization are now concentrating on university-led innovations and how these innovations are transferred into entrepreneurship and other value-adding activities (Braunerhjelm, 2007). Most recently, to ease the pressure of commercialization, many universities in many countries have started a technology transfer office (TTO) or technology licensing office (TLO). Governments have also taken initiative to foster university-led research (Rasmussen, 2008).

Proposition 1: Since IP is an economic tool for governments, universities, and industries, it is of prime importance to study the status and the possible economic benefits from the Malaysian university-led IP industry.

Commercialization of IP is about effectively transferring intellectual properties to greater use by an industry. Therefore, in a generic sense, a large number of industries (demand side) and institutions (supply side) will be involved in the process, and a wide variety of factors relating to every stakeholder will affect the success of IP commercialization (Braunerhjelm, 2007; Debackere and Veugelers, 2005; Feldman et al., 2002). The objective of this study was to explore the challenges and factors affecting IP commercialization of university-led innovations in the Malaysian context. Therefore, the primary concern of this study was to identify factors related to three stakeholders: universities, industries, and the government.

Proposition 2: Emerging countries face various challenges to successful IP commercialization. This study sought to explain the IP commercialization process in the Malaysian context, from innovation creation in universities to innovation diffusion in industries.

The major part of IP literature has highlighted legal issues, especially concerning the role of government policies and laws, that impact IP commercialization. Instead of working as the primary manipulator, the government is now working as a facilitator for both industries and universities (Rasmussen, 2008). Governments have a significant role to play in arranging financial mechanisms and nonfinancial instruments. These are unique to industries (such as biotechnology) and cultures (Chang et al., 2005). So the role of government is now mostly to influence the factors related to industries and universities, to facilitate effective IP commercialization, and to evaluate countrywide development.

Proposition 3.1: The IP laws and initiatives taken up by the government of Malaysia play an essential role in IP commercialization in Malaysian universities.

Innovation diffusion theory (Rogers, 1962) captures the demand side of an IP, meaning how innovation is transmitted through different customer groups in an industry. The success of IP commercialization is affected by competitiveness in the industry (Bhaduri nee Chakraborty and Mathew, 2003), governance of the IP mechanism in public-funded research (Geuna and Muscio, 2009), and support mechanisms for entrepreneurship (Hearn, Cunningham, et al., 2004). Government has a two-sided impact since it works as a supplier of funds for universities and also creates a policy environment for the industries that fosters commercialization.

Proposition 3.2: Industry works as the demand side of innovation, which plays a vital role in deciding what to innovate. This study proposed that various industry-related factors can influence IP commercialization in Malaysia.

Finally, institutional infrastructures also play a role, especially in terms of university research infrastructure, policies and organizational standards (Argyres and Liebeskind, 1998), productive technology transfer offices set up at universities (Debackere and Veugelers, 2005), faculty entrepreneurship, and interdisciplinary cooperation (Fakheri and Fazel, 2006). These are among the essential factors affecting university IP commercialization.

Proposition 3.3: Universities are considered the source of innovation. Most IP innovation is now university-led. Better university infrastructure and more supportive environments would foster innovation, and therefore encourage IP commercialization as well. As highlighted earlier, university-related factors can affect IP commercialization in Malaysia.

Over the past 25 years, a massive initiative has been undertaken by universities, industries, and governments to facilitate proper creation and dissemination of innovation in support of socioeconomic development (Fakheri and Fazel, 2006). Demand for innovations from industries and policy tools of the government have been the main drivers of much of the innovation effort (Bhaduri nee Chakraborty and Mathew, 2003; Borg, 2001; Chang et al., 2005; Debackere and Veugelers, 2005; Feldman et al., 2002).

Geuna and Muscio (2009) point out that, in order for the commercialization of university-led IP innovation to succeed, it requires an integrated plan involving the government, universities, and industries. The two forces of influences from the government, on both the supply and the demand sides, are assumed to have some interconnection since the government will have a long-term holistic economic plan and IP is a part of it. Finally, effective management of the university research infrastructure and industry competitiveness will foster IP commercialization, which is in turn again affected by government policies.

3 CONCLUSION

In conclusion, higher dependency on the government, lack of mature industry demands, and lack of adequate university research infrastructures mainly contribute to this semi-structured process. However, since the contribution of the government is much higher under the existing procedure, more policy implications can be expected at the national level, which is evolving similarly to both US and non-US styles (Furman, Porter, and Stern, 2002).

Comparison of two universities reveals similar issues, and it has opened up a solution for enterprises by changing the organization through a mix of innovation, technology, training, and demonstrable skills (Masbuqin and Sharif, 2019). Because of their varied approaches, research expertise, scope of research, and research experience, UTeM and USM are different. If the university scope matched a national innovation strategy, there would be greater possibility of getting accessible grants and innovating further. These are some of the benefits enjoyed by USM compared to UTeM.

ACKNOWLEDGMENT

The author would like to thank Universiti Teknikal Malaysia Melaka (UTeM) and the Centre for Technopreneurship Development (CTED) for their support in obtaining materials for this study and research.

REFERENCES

Alias, N. K. et al. (2018). The impact of knowledge management towards employees' job satisfaction. *International Journal of Academic Research in Business and the Social Sciences*, 8(9).

Borg, E. A. (2001). Knowledge, information and intellectual property: Implications for marketing relationships. *Technovation*, 21(8), pp. 515–524.

Debackere, K. and Veugelers, R. (2005). The role of academic technology transfer organizations in improving industry science links. *Research Policy*, 34(3), pp. 321–342.

Feldman, M. et al. (2002). Equity and the technology transfer strategies of American research universities. *Management Science*, 48(1), pp. 105–121.

Furman, J. L., Porter, M. E., and Stern, S. (2002). The determinants of national innovative capacity. *Research Policy*, 31(6), pp. 899–933.

Geuna, A. and Muscio, A. (2009). The governance of university knowledge transfer: A critical review of the literature. *Minerva*, 47(1), pp. 93–114.

Hamzah, Z. (2006). *Intellectual Property Law & Strategy*. Sweet & Maxwell Asia.

Jaiya, G. S. (2008). *Intellectual Property Management and Commercialization of New Products*. Stockholm: World Intellectual Property Organization.

Masbuqin, I. and Sharif, S. (2019). A study of the effectiveness of e-commerce platforms among small-medium enterprises (SME) in the postnatal care services industry.

Mat Isa, A. (2009). *Records Management and the Accountability of Governance*. Glasgow: University of Glasgow.

Mat Isa, A. et al. (2016). Knowledge sharing behaviour in libraries: A case study of Raja Tun Uda Library, Selangor, Malaysia.

Sharif, S., Ahamat, A., et al. (2018). University intellectual property commercialization: A critical review of the literature. *Turkish Online Journal of Design Art and Communication*, 8, pp. 874–886.

Sharif, S., Nizam, N., et al. (2018). Role of values and competencies in university intellectual property commercialization: A critical review. *Turkish Online Journal of Design Art and Communication*, 8, pp. 887–904.

Whittaker, J. B. (2003). Strategy and performance management in the government (white paper, p. 3). Mountain View, CA: Pilot Software.

World Intellectual Property Organization (WIPO) (1967). *World Intellectual Property Organization Convention*. Stockholm: World Intellectual Property Organization, p. 2.

Woodward, C. (2004). Valuation of intellectual property. Pricewaterhouse Coopers. http://mat

Facing Global Digital Revolution – Nirmala Arum Janie, Dwi Mulyaningsih & Wahyu Rachmawati (eds)
 ISBN 978-0-367-33912-8

Fraud drifts in the Grobogan Regency government, Indonesia

Yulianti Yulianti, Fariz Nur Hidayat & Dyah Nirmala Arum Janie
Universitas Semarang, Semarang, Indonesia

ABSTRACT: This research aimed to explore the perception of employees in local government agencies, enforcement of regulations, effectiveness of internal control, unethical behavior, compliance compensation, and the local government sector's commitment to fighting fraud. A purposive sampling technique was used. The sample in this research consisted of government agency employees in Grobogan Regency. Through questionnaires, 77 independent variables such as asymmetry of information, enforcement of regulations, effectiveness of internal control, unethical behavior, compliance compensation, and organizational commitment to fighting fraud in local government sectors were examined using multiple linear regression analysis and SPSS 20.0. The results of this research indicated that unethical behaviors have meant a significant inflation of fraud in the local government sector. Meanwhile, asymmetry of information, enforcement of regulations, effectiveness of internal control, compliance compensation, and organizational commitment also have highly important impacts on fraud in the local government sector.

Keywords: fraud in the local government sector

1 INTRODUCTION

The increasing demands of the community for the implementation of clean, fair, transparent, and accountable governance should accompany increased supervision of the financial management of a region. With this supervision, financial managers can provide an independent assessment of the central and regional governments to test and manage regional finance. The current government sector is very alarming because both the central and regional governments have seen much fraud.

Grobogan Regency is one of the corruption-affected districts in Central Java. Several corruption cases before the courts have seized public attention:

1. The corruption case of village fund allocation from the Grobogan regional budget in 2014 involving the head of the Tegal village of the Brati Grobogan subdistrict;
2. The E-KTP corruption case of Rp. 1.7 million, in which three of the defendants were sentenced to one and a half years in prison each;
3. The corruption case regarding the construction of the elephant road, which includes a package of Rp. 12 billion that involves the lead contractor (CV Sukma Jaya) and the commitment-making officer of the Public Works Agency of Bina Marga, each sentenced to one and a half years in prison; and
4. The corruption case for DPRD proper car maintenance (Rp. 1.8 million) involving three former council secretaries and the Grobogan DPRD chairperson.

Seeing the increasingly widespread level of corruption in Central Java, especially in Grobogan Regency, researchers tried to select research objects in Grobogan Regency.

Scholars have already carried out much research on fraud. This research has been conducted by exploring the perceptions of employees in government sector agencies to find trends of fraud in the government sector. This research was a replication and development of previous research performed by Mustika, Hastuti, and Heriningsih (2016). The reason for replicating the research was that some of the previous studies show different results, and not all of those relate to fraud.

The difference between this research and earlier research (Mustika et al., 2016) is that this study added one variable, namely organizational commitment. We based the addition of this variable on Chandra (2015). The second difference lies in the object of research; Mustika et al.'s (2016) research was conducted in Way Kanan Regency, Lampung, while this study concentrated on Grobogan Regency, Central Java. The public sector in Grobogan has undertaken much fraud, so further research is necessary because Grobogan is expected to be more transparent and open to the public through the use of financial statements.

2 LITERATURE REVIEW

According to Arfan (2011), *perception* is how people see or interpret events, objects, and other people. People act based on their perception by ignoring whether that perception reflects actual reality. Fraud can occur in all parts of an organization, ranging from the highest ranks to the lowest. Adinda (2015) states that *fraud* is a misrepresentation of material facts that is false, deliberate, and careless and that is believed and acted on so as to bring damage to a victim. Statement of Auditing Standard No. 99 (2002), presented in Ahriati, Basuki, and Widiastuty (2016), defines fraud as an intentional act to produce material misstatements in the financial statements that are the subject of an audit. Material misstatement in financial statements will mislead stakeholders or users of the financial statements because the information contained in them does not reflect the actual condition of the organization.

According to Scott (2009), as cited in Mustika et al. (2016), *asymmetry of information* is an imbalance in information acquisition between the trustee (local government/executive) as an information provider/agent and the trustee (community, people's representatives, legislature) as information users/principals. In this study, the asymmetry of information in question was asymmetry that occurs between government employees in the financial sector as agents and society as principals. According to Huda (2012), as quoted in Najah (2013), *regulation* is a pattern that manages behavior. So rules can be a bond, and they must be obeyed by all members of an organization so that the organization can run effectively and efficiently. IAPI (2011: 319.2), as referenced in Agoes (2012), states that *internal control* is a process carried out by a board of commissioners, management, and other personnel entities designed to provide adequate confidence in achieving three objectives, namely financial report reliability, effectiveness and efficiency in operations, and compliance with applicable laws and regulations. *Unethical behavior* is activity that is not generally accepted per social norms, in connection with beneficial or harmful actions (Griffin and Ebert, 2006, cited in Mustika et al., 2016). In this study, the unethical behavior in question was the behavior of the head of the agency, the head of the finance department, and the financial staff members responsible for accounting and for the preparation of financial reports within the district government.

According to Giterarry Desler (1996), *compensation* is all forms of rewards that flow to employees, while T Hani Handoko (1995) states that it is everything employees receive as a reward for their work. Mathis (2001), as quoted in Najah (2013), contends that *organizational commitment* is a level of trust and acceptance of labor toward organizational goals, and a desire to remain within an organization. Organizational commitment is the mental attitude of an individual with a level of loyalty toward the organization where he or she works. Therefore, the hypotheses examined in this study were:

Ha1: Asymmetry of information affects fraud in the government sector.
Ha2: Enforcement of regulations affects fraud in the government sector.
Ha3: The effectiveness of internal control affects fraud in the government sector.
Ha4: Unethical behavior affects fraud in the government sector.
Ha5: Satisfactory compensation affects fraud in the government sector.
Ha6: Organizational commitment affects fraud in the government sector.

3 METHOD

The researchers used purposive sampling for this study. The sample in this research comprised employees of government agencies in Grobogan Regency. The data collection consisted of questionnaires on the influence of 77 independent variables such as asymmetry of information, enforcement of regulations, effectiveness of internal control, unethical behavior, compliance compensation, and organizational commitment to fighting fraud in the local government sector, and these variables were examined using multiple linear regression analysis and SPSS 20.0

4 RESULTS

The coefficient value of asymmetry of information on fraud was 0.124, and the significance was 0.269. Therefore, we can conclude that the asymmetry of information variable does not affect fraud, and that the higher the asymmetry of information, the lower the occurrence of fraud. The results of this study indicate that the existence of good morality will minimize the occurrence of fraud in the government sector because morality affects the behavior of an individual. Asymmetry of information that occurs between agents and principals is seen from internal monitoring activities carried out thoroughly in each government agency. The results of this study are in line with research from Mustika et al. (2016) and Ahriati et al. (2016), which indicates that asymmetry of information does not affect fraud. However, this research is not in line with previous research conducted by Mustikasari (2013), Najahingrum et al. (2013), Chandra (2015), and Adinda (2015), which shows that the higher the asymmetry of information, the higher the tendency for fraud.

The regulation Pengakan influence coefficient value against fraud is a significant 0.151 and has a significance of 0.351. Therefore we can conclude that the influencing variable of regulatory influence does not affect fraud. We reject this hypothesis because the enforcement of this rule cannot detect fraud. The problem originates with and leads to "the man behind the gun." Whatever rules and procedures are created, they are strongly influenced by humans who hold power (Alpinista, 2013, cited in Najah, 2013). The uprightness of the rules in an entity depends on the morality of the authorized official. The results of this study are supported by Zulkarnain (2013) and Adinda (2015), which state that enforcement of regulations does not have a significant effect on fraud. However, this research contradicts Mustikasari (2013), Najah (2013), Chandra (2015), and Adinda (2015), which find that regulatory enforcement affects government fraud. It means that the higher the perception of employees regarding the enforcement of regulations in a government agency, the lower the occurrence of fraud in the government sector.

The coefficient of the effectiveness of internal control against fraud is equal to −0.257 and has a significance of 0.101. Therefore we can conclude that this variable does not affect fraud. In this study, agency employees complied with procedures and systems per the flow of tasks and responsibilities. The lack of support for this hypothesis proves that collusion in an agency where individuals act together can simultaneously cover fraud so that internal control cannot detect it (Mustika et al., 2016). This research is in line with Mustika et al. (2016), which concludes that the effectiveness of internal control does not affect fraud. However, this research is not in line with Mustikasari (2013), Najah (2013), Chandra (2015), and Adinda (2015), which argue that the effectiveness of internal control influences fraud in the government sector.

The coefficient value of unethical behavior on fraud is 0.329, with a significance of 0.041. Therefore we can conclude that this variable influences fraud. In other words, the higher the unethical behavior, the higher the likelihood of cheating, and the existence of unethical behavior will increase the chances of fraud. The results of this study indicate that morality affects the behavior of an individual. In an organization or agency represented by management, management morality can influence employees' actions. So in a company or an agency, management morality influences unethical behavior carried out by the agency. This is in line with research conducted by Faisal (2013), Zulkarnain (2013), and Mustika et al. (2016) which states that the lower unethical behavior by any employee within an agency will get low fraud (fraud) that might happen.

The coefficient of influence of conformity compensation on fraud is approximately −0.278 with a significance of 0.080. Therefore we can conclude that the regulatory influence variable does not affect fraud. This research shows that compensation does not necessarily suppress fraud. This may be because government officials provide inadequate compensation. This study was supported by Adinda (2015), and Mustika et al. (2016) show that compensation does not affect fraudulent tendencies. This study contradicted, however, Mustikasari (2013) and Chandra (2015), which find that compensation affects the tendency to commit fraud.

The coefficient value of organizational commitment on fraud is equal to −0.052 with a significance of 0.679. Therefore we can conclude that this variable does not affect fraud. This research shows that every government institution has a high organizational commitment; we cannot suppress the tendency of accounting fraud. We cannot commit verbally without high loyalty to the organization where we work. The results of this study are supported by Chandra (2015), which finds that organizational commitment does not affect fraud. So the lower the organizational commitment, the higher the level of fraud. The results of this study contradict Mustikasari (2013), Najah (2013), and Adinda (2015), however.

5 CONCLUSION

The limitation of this study is that the relatively small adjusted R^2 value is only 40; 6% of the variables used in this study affect the tendency of fraud by only 40.6%. The remaining 59.4% is influenced by other variables not tested in this study. Future studies can add independent variables and expand research objects in order to obtain more samples. Besides, the next researcher can also deepen the indicators that can influence fraudulent tendencies in regional government. Then he or she can interview respondents to get more accurate results while also exploring the elements that cause fraud by using financial statements.

REFERENCES

Adinda, Y. M. (2015). Faktor yang mempengaruhi terjadinya kecurangan (fraud) di sektor pemerintahan kabupaten klaten. *Accounting Analysis Journal*, 1(2), pp. 1–6.

Agoes, S. (2012). *Auditing: Petunjuk Praktis Pemeriksaan Akuntan Oleh Akuntan Publik*. Jakarta: Salemba Empat.

Ahriati, D., Basuki, P., and Widiastuty, E. (2016). Analisis pengaruh sistem pengendalian internal, asimetri informasi, perilaku tidak etis dan kesesuaian kompensasi terhadap kecenderungan kecurangan akuntansi pada Pemerintah Daerah Kabupaten Lombok Timur. *InFestasi*, 11(1), pp. 41–55.

Arfan, L. I. (2011). *Akuntansi Keperilakuan, Cetakan Kedua*. Jakarta: Salemba Empat.

Chandra, D. P. (2015). Determinan terjadinya kecenderungan kecurangan akuntansi (fraud) pada dinas pemerintah kabupaten grobogan. Universitas Negeri Semarang.

Faisal, M. (2013). Analisis fraud di sektor pemerintahan kabupaten kudus. *Accounting Analysis Journal*, 2(1).

Mustika, D., Hastuti, S., and Heriningsih, S. (2016). *Analisis Faktor-Faktor Yang Mempengaruhi Kecenderungan Kecurangan (Fraud): Persepsi Pegawai Dinas Kabupaten Way Kanan Lampung*. Lampung: Simposium Nasional Akuntansi XIX.

Najah, A. F. (2013). Faktor-faktor yang mempengaruhi fraud: Persepsi pegawai Dinas Provinsi DIY. *Accounting Analysis Journal*, 2(3).

Zulkarnain, R. M. (2013). Analisis faktor yang mempengaruhi terjadinya fraud pada dinas Kota Surakarta. *Accounting Analysis Journal*, 2(2), pp. 125–130.

Facing Global Digital Revolution – Nirmala Arum Janie, Dwi Mulyaningsih & Wahyu Rachmawati (eds)
 ISBN 978-0-367-33912-8

Corporate environmental disclosure as a form of social responsibility in the annual report

Ardiani Ika Sulistyawati, Febriyani Intan Permata & Dyah Nirmala Arum Janie
Universitas Semarang, Semarang, Indonesia

ABSTRACT: This study aims to examine the influence of managerial ownership, company size, and financial performance on the level of disclosure and reporting of environmental information in the annual report of environmentally sensitive companies registered in PROPER in 2011-2015. This study uses the Purposive Sampling method. The population in this study is all companies listed in the IDX as well as registered as PROPER participants for 2011-2015 period. Moreover, the sample used in this study consists of 38 companies listed in Indonesia Stock Exchange data from 2011-2015. The results of this study indicate that managerial ownership has a significant effect on corporate environmental disclosure. Company size has a significant positive effect on corporate environmental disclosure. And finally, financial performance has a significant effect on corporate environmental disclosure.

Keywords: managerial ownership, company size, financial performance and corporate environmental disclosure

1 INTRODUCTION

Indonesia is a rich country in terms of the potential it has from the extraordinary natural resources, ranging from marine and terrestrial wealth to those contained in it. The country has the world's third largest tropical rainforest. However, behind this abundant natural wealth, Indonesia experiences various environmental problems (Diana and Amalia, 2015). It ranges from natural disasters to climate change to environmental damage. Environmental damage caused by company's activities is increasingly widespread. This is an evidence of how companies have not been optimally responsible for the environment. Despite the numerous regulations and standards on environmental management, their implementations are still far from expectation even until recently.

In recent years, Indonesia has experienced an increase in environmental pollution problems (Heriningsih and Saputri, 2015). We can see this environmental pollution from various disasters that have occurred lately, such as flash floods in certain areas. Environmental problems are a serious concern for everyone, including consumers, investors and the government.

In general, investors are more interested in companies that implement a proper environmental management and do not ignore environmental pollution problems (Ja'far and Arifah, 2006).For the purpose of showing their reputation, credibility and value added for the company in the eyes of their stakeholders, companies are encouraged to disclose their social responsibility towards the environment in their annual reports (Eipstein and Freedman, 1994 in Suhardjanto and Permatasari, 2010). Financial accounting standards in Indonesia have not required companies to disclose environmental information (Suhardjanto, 2008). As a result, not many companies disclose their environmental activities (Anggraini, 2006).

The existence of industry amidst the society has an impact on people's lives. Economically, the existence of industry improves people's welfare by providing employment opportunities. To prevent the environmental problems from becoming more complex, the government has attempted to issue a legal umbrella, namely Law of Limited Liability Company No. 40 Article 74 Year 2007. The article explains that any company running its business activities in the field of and or related to natural resources must carry out their social and environmental responsibilities.

The government, through the Ministry of Environment, also releases a company performance appraisal program in environmental management known as PROPER. With this PROPER rating, companies whose activities have to some extent an impact on the environment are expected to be more transparent in disclosing and reporting their environmental information (Diana and Amalia, 2015).

Corporate Environmental Disclosure is the disclosure of information relating to the environment in a company's annual report. The indicator used in measuring environmental disclosures in this study is the environmental disclosure standard at the Global Reporting Initiative (GRI). The current Corporate Environmental Disclosure is still voluntary; this leads the government, the company and the community to blame each other and avoid for being held accountable for the occurring environmental damages. Corporate Environmental Disclosure has also received much criticism, not only in Indonesia but also throughout the world.

The financial statements selected as the source of data for this study are those during 2011-2015 period to find out such information as company size, managerial ownership, and ROA, as they are used as indicators of financial performance. The Government through the Ministry of Environment also releases a company performance appraisal program in environmental management known as PROPER. PROPER is a form of transparency and involvement of the community in environmental management, where the results of supervision through PROPER, will be delivered openly to the public.

The research conducted by Yahya (2007) finds that profitability harms the disclosure of environmental information. Companies are more likely to be profit-oriented, where when these companies gain high profits, they think it is unnecessary for them to disclose their social responsibility since they have been financially successful. These results contradict the research conducted by Patten (1992) and Tarmizi (2012) who find stating that profitability has a positive effect on disclosure of environmental information.

Environmental problems are a serious concern for everyone, including consumers, investors, and the government. In general, investors are more interested in companies that implement proper environmental management and do not ignore environmental pollution problems (Ja'far and Arifah, 2006). Since not many Environmental Disclosure studies have been carried out in Indonesia, yet on the other hand the environmental pollution is getting worse in the country, research on environmental accounting in particular environmental disclosure is fundamental.

Dian's (2009) research results show that company size has a positive effect on environmental disclosure. The results of Dian's (2009) study confirm the results of Sembiring's (2005) study, who finds that company size has a significant effect on corporate social responsibility disclosure. Some studies that have shown the positive relationship between these two variables include Suhardjanto & Afni (2009) and Hanifa & Cooke, 2005 (in Suhardjanto and Choiriyah, 2010).

2 LITERATURE REVIEW

Corporate environmental disclosure is a disclosure made by a company to stakeholders as a report of environmental activities undertaken by this company. Al Tuwaijri (2004) defines environmental disclosure as a collection of information related to environmental management activities by companies in the past, present and future. We can find this information in qualitative statements, assertions or quantitative facts, forms of financial statements or footnotes. The field of environmental disclosure includes expenses or operating costs for facilities of pollution control equipment in the past and present.

Tjeleni (2013) states that managerial ownership is a situation where the manager owns the company's shares, or in other words, the manager is also a shareholder. This presence of managerial ownership shows the existence of a dual role of a manager, i.e. the manager also acting as a shareholder.

Managerial ownership means a situation where shareholders also play the role of owners of the company and are engaged in the management and actively participate in making the company's decisions (Diana and Amalia, 2015). From these definitions, we can interpret managerial ownership as the proportion of ordinary shares owned by management. Management that owns company shares will certainly align its interests with those of its shareholders. Meanwhile, when managers do not have company shares, there is a possibility that they are only concerned with their interests. Managerial ownership is the separator of ownership between the outsider and the insider.

In terms of company size, large-scale companies are generally easier to access loans than smaller ones because it is related to the level of creditor trust in large companies. Large companies also tend to be more diversified and more resistant to the bankruptcy risk (Najmudin, 2011).

Firm size relates to agency theory which states that an attempt is carried out to avoid conflicts that occur between agents and principals. Company size will affect all items of the company, including the number of employees, the amount of production, company income and so forth.

According to the Indonesian Association of Chartered Accountants (2007), profitability is the company's ability to manage and control the resources it has. We measure financial performance in this study through profitability ratios. We define profitability as the company's ability to generate profits to increase shareholder value. Profitability is the ability of a company to earn profits through sales, total activities, and its capital. The level of profitability will be measured using financial ratios, namely Return on Assets (ROA). ROA is the ability of capital invested in all assets to generate profits for all investors, both bondholders and shareholders (Riyanto, 2013).

ROA is a financial indicator that describes the company's ability to generate profits on the total assets of the company. The higher the ratio, the better, because company management can generate the best possible return on the assets they own (Putri and Cristiawan, 2014).

3 METHOD

The object in this study is all PROPER companies listed in Indonesia Stock Exchange for 2011-2015 period. The population in this study is all companies listed in Indonesia Stock Exchange (IDX) and at the same time registered as PROPER participants for five years from 2011 to 2015. The sampling unit in this study is a company registered in PROPER and their financial statements are taken through Indonesia Stock Exchange for 2011-2015 period.

The population in this study is manufacturing companies listed in Indonesia Stock Exchange from 2011-2015. The sample is taken using Purposive sampling, i.e. the population used as the research sample is those that meets the criteria of a particular sample. The criteria for sampling to be examined are as follows:

Using the Pooled Time Series method, then from the data above, the companies used as the sample in this study for 2011-2015 period are 39 x 5, making a total of 195 companies. Below is the operating definitions of the variables we use.

Table 1. Operating definition of variables.

Variables	Definition	Indicator	Research
Managerial ownership	The management members have company shares.	The number of shares owned by management in a company.	Diana Oktafianti & Amalia Rizki 2011- 2013
Company Size	Describing the size of a company as measured through market capitalization, total capital used, total assets to total sales obtained (Yahya, 2007)	LN = total assets	Diana Oktafianti & Amalia Rizki 2011- 2013
Financial Performance	The company's ability to manage and control the resources it has.	Profitability ratio, namely ROA.	Diana Oktafianti & Amalia Rizki 2011- 2013
Corporate Environmental disclosure	Disclosure of information related to the environment in the annual report.	Environmental disclosure standards at the Global Reporting Initiative (GRI)	Diana Oktafianti & Amalia Rizki 2011- 2013

4 RESULTS

After going through a series of standard assumption tests, the analysis tests using multiple linear regression yield the following results:

1. The constant is positive and significant. It means that if managerial ownership, firm size, company performance is assumed to be constant, then the value of corporate environmental disclosure potentially increases significantly.
2. The managerial ownership regression coefficient value is positive and significant. It means that, assuming all independent variables are constant, then if the managerial ownership increases, then the corporate environmental disclosure also has the potential to increase significantly.
3. The regression coefficient of company size is positive and significant. It means that, assuming all the independent variables are constant, then if the size of the company increases, the corporate environmental disclosure also has the potential to increase significantly.
4. The regression coefficient value of company performance is positive and significant. It means that, assuming all independent variables are constant, then if the company performance increases, the corporate environmental disclosure also has the potential to increase significantly

The Adjusted R^2 value is 72.5%. It means the ability of independent variables to explain the dependent variable is about 72.5%, while other variables outside the model explain the remaining 27.5%. The P-value and T-value show significant numbers. Therefore, all variables in this study have a positive and significant effect on corporate environmental disclosure.

5 LIMITATIONS

This research has some weaknesses, one of which is a minimal number of ROA. For future research, we suggest to add the variables, lengthen the observation periods, and increase the number of samples.

REFERENCES

Diana, O. and Amalia, R. (2015) 'Kepemilikan Manajerial, Ukuran Perusahaan dan Kinerja Keuangan terhadap Corporate Environmental Disclosure', *Simposium Nasional Akuntansi 18*.

Heriningsih, S. and Saputri, N. (2015) 'Pengaruh Corporate Social Responsibility Disclosure dan Environmental Performance Terhadap Economic Performance pada Perusahaan Manufaktur yang Terdaftar di Bursa Efek Indonesia', *Jurnal Ekonomi Dan Bisnis*, 10(1).

Ja'far, M. and Arifah, D. A. (2006) 'Pengaruh Dorongan Manajemen Lingkungan, Manajemen Lingkungan Proaktif dan Kinerja Lingkungan terhadap Public Environmental Reporting', *Simposium Nasional Akuntansi*, 9, pp. 23–25.

Suhardjanto, D. and Choiriyah, U. (2010) 'Information GAP: Demand supply environmental disclosure di Indonesia', *Jurnal Keuangan dan Perbankan*, 14(1), pp. 36–51.

Suhardjanto, D. and Permatasari, N. D. (2010) 'Pengaruh corporate governance, etnis, dan latar belakang pendidikan terhadap environmental disclosure: Studi empiris pada perusahaan listing di Bursa Efek Indonesia', *Kinerja*, 14(2), pp. 151–164.

Undang-Undang No.40 Pasal 74 Tahun. (2007) 'Perseroan Yang Menjalankan Kegiatan Usaha Dibidang Sumber Daya Alam Wajib Melaksanakan Tanggung Jawab Sosial'.

Undang-Undangan No.25 Tahun. (2007) 'Penanaman Modal Terkait Dengan Perusahaan Yang Terdaftar Di Pasar Modal'.

Facing Global Digital Revolution – Nirmala Arum Janie,
Dwi Mulyaningsih & Wahyu Rachmawati (eds)
© 2020 Taylor & Francis Group, London, ISBN 978-0-367-33912-8

Efforts to improve the performance of micro, small and medium business companies

Paulus Wardoyo & Endang Rusdianti
Universitas Semarang, Semarang, Indonesia

ABSTRACT: Micro, Small and Medium Enterprises are businesses that are mostly in business in Indonesia. Its existence is regulated by Law number 20 of 2008. There are many studies of MSMEs, but there are still many left behind the controversy over the results of the research. This study used a sample of 100 respondents and selected using purposive. Data collection is done using questionnaires and analyzed using structural equations. Research proves that the ability to access information and competency in responding to the market is an antecedent of entrepreneurial orientation, and is useful for improving company performance.

Keywords: Access to Information, Responding to Markets, Company Performance

1 INTRODUCTION

The Law of the Republic of Indonesia number 20 of 2008 describes the classification of Micro, Small, and Medium. Referring to several previous studies as well as those conducted by (Lukiastuti, 2010; Tang *et al.*, 2007; Wiklund & Shepherd, 2003) and studies conducted by (Murniningsih, 2015.) show similar results, where entrepreneurial orientation has a positive influence on company performance. However, the results of these studies are very contrary to the studies conducted by (Baker and Sinkula, 2009) as well as (Awang, Asghar and Subari, 2010) which both conclude that entrepreneurial orientation does not affect company performance.

Performance measurement in this way usually uses qualitative and quantitative methods (Yildiz and Karakas, 2012 quoted by Murniningsih, 2015). Whereas Saraf et al (2007) cited by Murniningsih (2015) view sharing information with partners as something important because it will be able to obtain competitive advantage from the sources of knowledge. The market response is the competence of the company, if done well, will be able to create a competitive advantage (Deshpande and Webster Jr, 1989; Kohli & Jaworski, 1990; Svensson, 2004).

The results of the study Wiklund and Shepherd (2005) concluded that companies that are more entrepreneurial oriented have better performance. The purpose of this study is to build a model that can be used to improve MSME business performance. Based on the previous description, the research hypothesis reads as follows:

H1: The higher the ability to access information, the higher the entrepreneurial orientation.
H2: The higher the competency responds to the market, the higher the entrepreneurial orientation.
H3: The higher the ability to access information the higher the company's performance.
H4: The higher the entrepreneurial orientation the higher the company's performance
H5: The higher the competency responds to the market that is owned by a company, the higher the company's performance.

2 RESEARCH METHODS

This study uses primary data. This type of data is obtained directly from the questionnaire to MSME business actors in Semarang City. In this study the population used was MSME entrepreneurs assisted by the Kop & UMKM Service in Semarang City with 3,066 numbers, the sampling method used purposive sampling, with 5 times the number of indicators. The number of indicators in this study was 15 indicators, so the number of samples was 75 respondents. Analysis techniques use structural equations; data processing is carried out using the Smart PLS program version 1.0

3 RESULTS AND DISCUSSION

The results of processing at the initial stage, found a loading factor of 0.584 on the OW1 indicator, so it must be removed from the model. Then the model is reprocessed, the results show that the effect of the KMI variable on the KP value is greater than 1, so it is necessary to verify the model, by issuing the KMI4 indicator. The final results show a better model, as in the figure below

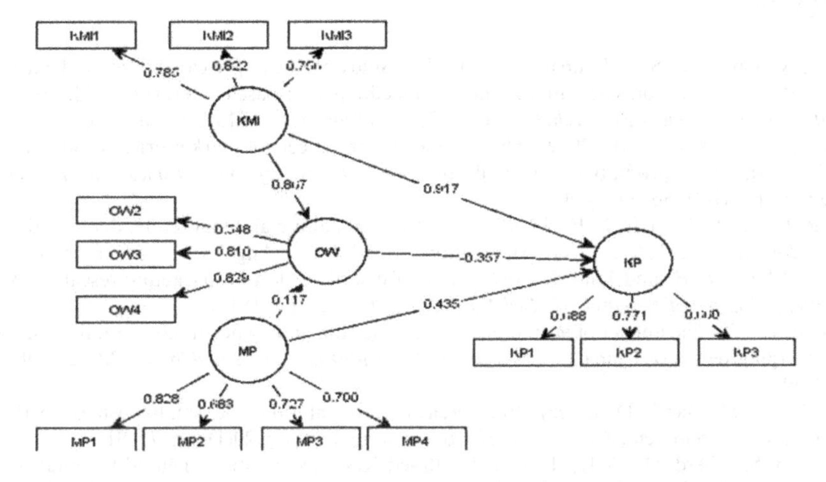

Figure 1. Full model.

Next, referring to the picture above, and the opinion of Chin (1998), it can be concluded that all of these indicators are acceptable. The path coefficient results structural and indicators and their significance values are shown in the following table:

Table 1. Data processing results.

	original sample estimate	mean of subsamples	Standard deviation	T-Statistic
KMI -> OW	0.867	0.858	0.04	21.462
MP -> OW	0.117	0.131	0.071	1.651
KMI -> KP	0.917	0.912	0.236	3.892
MP -> KP	0.435	0.437	0.096	4.511
OW -> KP	-0.357	-0.357	0.253	1.41

Note: KMI = Ability to Access Information, OW = Entrepreneurial Orientation, MP = Market Responding Competence, KP Company Performance

Refer to the table above, see that hypotheses 2 and 4 in this study were rejected, however, both of these hypotheses were significant at an alpha of 10% (Hulland, Chow and Lam, 1996), while H1, H3 and H 5 were accepted. This study successfully proved that the ability to access information and competency in responding to the market had an effect on entrepreneurial orientation. Furthermore, both the ability to access information, market competency and entrepreneurial orientation can influence company performance.

4 CONCLUSION

The business actors of MSMEs in improving the business performance that they are engaged in, are not enough to just have an entrepreneurial orientation but need to be supported also by the ability to access market information and competency in responding to the market. However, the ability to access market information has a more important influence than market competency. The existence of two hypotheses that are significant at 10% alpha, is a weakness of this study. Future studies should be done by increasing the number of respondents or focusing more on certain industry sectors.

REFERENCES

Awang, A., Asghar, A. R. S. and Subari, K. A. (2010) 'Study of distinctive capabilities and entrepreneurial orientation on return on sales among small and medium agro-based enterprises (SMAEs) in Malaysia', *International Business Research*. Canadian Center of Science and Education, 3(2), p. 34.

Baker, W. E. and Sinkula, J. M. (2009) 'The complementary effects of market orientation and entrepreneurial orientation on profitability in small businesses', *Journal of small business management*. Wiley Online Library, 47(4), pp. 443–464.

Deshpande, R. and Webster Jr, F. E. (1989) 'Organizational culture and marketing: defining the research agenda', *Journal of marketing*. SAGE Publications Sage CA: Los Angeles, CA, 53(1), pp. 3–15.

Hulland, J., Chow, Y. H. and Lam, S. (1996) 'Use of causal models in marketing research: A review', *International Journal of Research in Marketing*. Elsevier, 13(2), pp. 181–197.

Tang, J. *et al.* (2007) 'The impact of entrepreneurial orientation and ownership type on firm performance in the emerging region of China', *Journal of developmental Entrepreneurship*. World Scientific, 12(04), pp. 383–397.

Wiklund, J. and Shepherd, D. (2005) 'Entrepreneurial orientation and small business performance: a configurational approach', *Journal of business venturing*. Elsevier, 20(1), pp. 71–91.

Wiklund, J., & Shepherd, D. (2003) 'Knowledge Based Resources, Entrepreneurial Orientation and the Performance of Small and Medium Size Businesses', *Strategic Management Journal*, 24, 1307–1314.

Facing Global Digital Revolution – Nirmala Arum Janie, Dwi Mulyaningsih & Wahyu Rachmawati (eds)
© 2020 Taylor & Francis Group, London, ISBN 978-0-367-33912-8

Determinants of social budget preferences in the village government income and expenditure budgeting

Dian Indudewi, Febrina Nafasati & Abdul Karim
Universitas Semarang, Semarang, Indonesia

ABSTRACT: The performance of the Village Government was highlighted when Indonesian Corruption Watch (ICW) announced that misuse of the budget was the highest factor in corruption throughout 2017. ICW data showed that there were 98 cases of corruption in village administrations and 102 village heads being suspected. The budget is still a soft land for corruption. This phenomenon is a trigger to find out the actual social preferences of budget actors in preparing budgets and achieving budget performance. Limited financial capacity encourages the establishment of priority scale of government affairs where the government is required to choose and delay others. Exchange processes thought to arise when governments prepare budget and set budget revenue and expenditure. This paper seeks to see from the aspect of social exchange, how social exchange theory can be used to explain the phenomenon of budget misuse that still occurs and is able to form indicators to measure social preferences.

1 INTRODUCTION

Issuance of Law Number 6 Year 2014 concerning Villages is the basis for the implementation of autonomy in the Village area. Autonomy means that the village has the authority to organize government affairs starting from planning, budgeting, implementation, and accountability. A set of rules related to village budgeting mechanisms are issued to regulate the implementation of village autonomy, specifically budgeting mechanisms such as Minister of Home Affairs Regulation Number 113 of 2014 concerning Village Financial Management, Government Regulation Number 60 of 2014 and renewed with Number 22 of 2015 concerning Village Funds from State Budget.

Although the mechanism the budget has been regulated by legislation, budget actors apparently still have a gap to play in the budgeting process and even cause delays in budget approval (Febrina and Isril, 2016; Parwati, Nyoman Budiasih and Astika, 2015; Subechan, Hanafi and Haryono, 2014; Riyanto, 2012; Hanida, 2016). The Supreme Audit Agency (BPK) found that out of 6,222 weaknesses in the internal control system in the regional government, 2,887 cases were caused by weaknesses in controlling the implementation of the revenue and expenditure budget. The results of BPK's examination are in line with the mapping carried out by ICW regarding cases of corruption in 2017. ICW revealed that out of 576 corruption cases, 98 cases were misuse of the budget by the village government so that there were 102 village heads who were suspected of corruption.

Research related to government budgets in Indonesia so far referring to proving the opportunistic behavior of budget actors by looking at budget spreads (Parwati, Nyoman Budiasih and Astika, 2015; Sularso, 2014; Abdullah, 2004). Sujaie (2013) explored more in the opportunistic behavior of budget actors and found that there were bargaining and soft negotiations in the preparation of the East Java Provincial Budget. Research by Febrina and Isril (2016); Hanida (2016); Abdullah (2004) found that there was a difference of interests between the executive and the legislature, resulting in the budget being late. Research outside Indonesia related to the budget also provides interesting results. Goncalves (2014) found that even though there was community involvement, increasing welfare had to be done gradually. Ozdemir, Johnson & Whittington (2016) study found that ideology plays a role in choosing program and activity preferences, and funding sources are an important consideration when making policy. The results of previous studies indicate budgetary actors have certain preferences in preparing budgets.

Preferences may appear during the process of selecting programs and activities that will be incorporated into the budget. Election of preferences can be determined by the social relationships that exist between organizational members and between organizations. Social relations are social exchanges between individuals or organizations that have the same value to achieve common goals (Emerson, 1976; Blau, 1964). The existence of social structures can create imbalances, so social values become a counterweight that can create social balance (Cook, 2015; Cook *et al.*, 2013; Lawler and Yoon, 1993; Cook, 1977; Emerson, 1976). These social values can be in the form of altruism (Zuckerman, Cheng and Nau, 2018; Murphy, Ackermann and Handgraaf, 2011; Cropanzano and Mitchell, 2005), trust (Blay, Douthit and Fulmer III, 2019; Zuckerman, Cheng and Nau, 2018), and fairness (Falk *et al.*, 2016; Cropanzano and Mitchell, 2005; Gould-Williams and Davies, 2005; Andreoni, Brown and Vesterlund, 2002).

2 LITERATURE REVIEW AND THEORETICAL BACKGROUND

The law of reciprocity (giving and receiving) forms the basis of all social relations (Göbel, Vogel and Weber, 2013). On the other hand, preference refers to choices/priorities the main thing that an individual does is through considerations basically tends to contain fulfillment individual needs. Therefore, when individuals are faced with social choices that are around them, then the individual has motivational considerations which ultimately drop the choices that provide incentives for him. The interrelationship interpreted as a form of interaction that is based on equality (equality) and beneficial to all parties who interact (mutuality) (Murphy, Ackermann and Handgraaf, 2011; Bandiera, Barankay and Rasul, 2005; Andreoni, Brown and Vesterlund, 2002). Equality and mutual interaction that is interwoven in organizations encourages individuals to do what the organization wants (Göbel, Vogel and Weber, 2013). Individuals will be loyal, have high organizational commitment, manage organizational resources well, obey the rules in the organization. Not only is the organization-member relationship harmonious, but there will also be mutual trust and mutual support in the relationships between members of the organization. Negotiation is a way of reaching mutual agreement (Lawler, 1992; Thompson, 1991; Schelling, 1980; Cook and Emerson, 1978).

The village government which consists of village heads and village officials as perpetrators as well as budget compilers certainly faces social and even political choices in preparing the Village Budget (APBDes). Village funds obtained from the central government must be used as well as possible in the form of programs and activities that can prosper the village community. Social preferences inherent in individuals can have an impact on the selection of programs and activities that enter the APBDes. Therefore, knowing the forming factors of social preferences inherent in the APBDes budget actors is an interesting thing. These forming factors can later be used as indicators to measure social preferences and their impact on village government decision making and performance.

3 METHOD

Data was obtained by distributing 120 questionnaires to APBDes budget compilers namely the village head and village officials in the Karangawen Sub-District Environment. Karangawen subdistrict was chosen because it had been assisted by the village financial system so that it was expected that village officials could fill out questionnaires honestly and as they were. Questionnaires that were returned and could be processed were only 81 questionnaires (response rate 67.5%). Questionnaire in the form of a list of statements as many as 33 items related to social values in preparing the budget. Factor formation is done by factor analysis test.

4 RESULT AND DISCUSSION

To perform a factor analysis test, several assumptions about factor analysis must be fulfilled. Distribution of respondent's answer data distribution with boxplot test shows even distribution in the median value range. The KMO value is 0.916 (> 0.5) and the Bartlett test shows a sig value of 0.00 (<0.05). The measure of sampling adequacy (MSA) was all above 0.5. These results indicate assumptions that have been fulfilled and can proceed to the factor analysis test .

Test for the formation of factor analysis is carried out through several stages. The first stage is the analysis of communalities where all extract values> 0.05 so that it can proceed to the next stage, namely the formation of factors. Based on the total initial eigenvalue fourth unknown factors that may be formed. Factor 1 consists of the variable matrix of APBDes alignments, assumptions of APBDes preparation, APBDes justice, proportional budget, community access, political aspects, pro-society, evaluation process, data updating, priority interests, and socio-cultural aspects. Factors consists of commitment, mutual trust, cooperation, uniformity of understanding, trust in APBDes, and ease of access. Factor 3 consists of community welfare, community needs, village needs, community interests, and mutual success. Factor 4 consists of material motivation, material justice, equivalence of reward, authority, and political pressure. Based on social exchange theory, factor 1 is an altruism variable (Zuckerman, Cheng and Nau, 2018; Murphy, Ackermann and Handgraaf, 2011; Cropanzano and Mitchell, 2005). Factor 2 is a trust variable (Blay, Douthit and Fulmer III, 2019; Falk *et al.*, 2016; Cropanzano and Mitchell, 2005). Factor 3 is a power dependence variable (Kölln, 2018; Molm, 2014). Forming factor 4 is equality variable (Murphy, Ackermann and Handgraaf, 2011; Bandiera, Barankay and Rasul, 2005; Andreoni, Brown and Vesterlund, 2002).

5 CONCLUSION

This study resulted in 4 factors forming social preferences that are in line with social exchange theory, where budgetary actors have social preferences in the form of altruism, trust, power dependence, and equality. This forming factor can be used as a dimension to measure social preferences. Although the correlation value between variable matrices is high, the ability of the forming factors in summarizing the 33 variable matrices is low. This can be due to the relatively small data and none of the village heads filling out the questionnaire. This is very unfortunate because the village head has the highest authority in the budgeting process . The results of answers tend to be in a neutral range. Related to the number of village heads who are suspected of corruption can be caused by a lack of knowledge of budgeting, so that the budget allocation and proportionality are not properly carried out. This is also supported by respondents' answers where only 2% of village officials have participated in budget training. For better research results, it is expected that further research can develop a research model based on social preferences, expand the research location, and get the village head as the respondent.

REFERENCES

Andreoni, J., Brown, P. M. and Vesterlund, L. (2002) 'What makes an allocation fair? Some experimental evidence', *Games and Economic Behavior*. Elsevier, 40(1), pp. 1–24.

Bandiera, O., Barankay, I. and Rasul, I. (2005) 'Social preferences and the response to incentives: Evidence from personnel data', *The Quarterly Journal of Economics*. MIT Press, 120(3), pp. 917–962.

Blay, A., Douthit, J. and Fulmer III, B. (2019) 'Why don't people lie? Negative affect intensity and preferences for honesty in budgetary reporting', *Management Accounting Research*. Elsevier, 42, pp. 56–65.
Cook, K. S. (1977) 'Exchange and power in networks of interorganizational relations', *The sociological quarterly*. Taylor & Francis, 18(1), pp. 62–82.
Cook, K. S. *et al.* (2013) 'Social exchange theory', in *Handbook of social psychology*. Springer, pp. 61–88.
Cook, K. S. (2015) 'Exchange: Social', *International Encyclopedia of the Social & Behavioral Sciences: Second Edition*. Elsevier, 8.
Cook, K. S. and Emerson, R. M. (1978) 'Power, equity and commitment in exchange networks', *American sociological review*. JSTOR, pp. 721–739.
Cropanzano, R. and Mitchell, M. S. (2005) 'Social exchange theory: An interdisciplinary review', *Journal of management*. Sage Publications Sage CA: Thousand Oaks, CA, 31(6), pp. 874–900.
Emerson, R. M. (1976) 'Social exchange theory', *Annual review of sociology*. Annual Reviews 4139 El Camino Way, PO Box 10139, Palo Alto, CA 94303-0139, USA, 2(1), pp. 335–362.
Falk, A. *et al.* (2016) 'The preference survey module: A validated instrument for measuring risk, time, and social preferences'. IZA Discussion Paper.
Febrina, R. and Isril (2016) 'Political Process Analysis Discussion and Determination of Regional Revenue and Expenditure Budget (APBD) of Riau Province Budget Year 2016'.
Government Regulation Number 60 of 2014 concerning Village Funds Sources From the State Revenue and Expenditure Budget.
Göbel, M., Vogel, R. and Weber, C. (2013) 'Management research on reciprocity: A review of the literature', *Business Research*. Springer, 6(1), pp. 34–53.
Gould-Williams, J. and Davies, F. (2005) 'Using social exchange theory to predict the effects of HRM practice on employee outcomes: An analysis of public sector workers', *Public management review*. Taylor & Francis, 7(1), pp. 1–24.
Hanida, R. P. (2016) 'Dinamika Penyusunan Anggaran Daerah: Kasus Proses Penetapan Program Dan Alokasi Anggaran Belanja Daerah di Kabupaten Sleman', *Jurnal Penelitian Politik*, 7(1), p. 15.
Indonesian Public Revised Financial Supervisory Agency (BPK-RI). (2018) 'Overview of Results of Examination for Semester I 2018.
Kölln, A.-K. (2018) 'Political sophistication affects how citizens' social policy preferences respond to the economy', *West European Politics*. Taylor & Francis, 41(1), pp. 196–217.
Lawler, E. J. (1992) 'Power processes in bargaining', *Sociological Quarterly*. Wiley Online Library, 33(1), pp. 17–34.
Lawler, E. J. and Yoon, J. (1993) 'Power and the emergence of commitment behavior in negotiated exchange', *American Sociological Review*. JSTOR, pp. 465–481.
Minister of Home Affairs Regulation Number 113 of 2014. Regarding Village Financial Management.
Molm, L. D. (2014) 'Experiments on exchange relations and exchange networks in sociology', in *Laboratory experiments in the social sciences*. Elsevier, pp. 199–224.
Murphy, R. O., Ackermann, K. A. and Handgraaf, M. (2011) 'Measuring social value orientation', *Judgment and Decision making*, 6(8), pp. 771–781.
Parwati, S. M., Nyoman Budiasih, I. G. A. and Astika, I. B. P. (2015) 'Opportunistic Behavior of Budgeters', *Accounting and Business Scientific Journal*, 10(2), pp. 127–133.
Schelling, T. C. (1980) *The strategy of conflict*. Harvard university press.
Subechan, S., Hanafi, I. and Haryono, B. S. (2014) 'Analisis Faktor-faktor Penyebab Keterlambatan Penetapan APBD Kabupaten Kudus', *WACANA, Jurnal Sosial dan Humaniora*, 17(1), pp. 35–46.
Sujaie, A. F. (2013) 'Oportunisme Perumus Kebijakan Anggaran dalam Penyusunan APBD Provinsi Jawa Timur Tahun 2013: Fenomena dalam Pelaksanaan Belanja Hibah'. Universitas Gadjah Mada.
The government ran Peratu No. 22 of 2015 on Amendment of Government Regulation No. 60 Year 2014 On The Village Fund Sourced From Budget Revenue and Expenditure.
Thompson, L. L. (1991) 'Information exchange in negotiation', *Journal of Experimental Social Psychology*. Elsevier, 27(2), pp. 161–179.
Undang-Undang Nomor 6 Tahun 2014 about the Village.
Zuckerman, I., Cheng, K.-L. and Nau, D. S. (2018) 'Modeling agent's preferences by its designer's social value orientation', *Journal of Experimental & Theoretical Artificial Intelligence*. Taylor & Francis, 30(2), pp. 257–277.

Facing Global Digital Revolution – Nirmala Arum Janie,
Dwi Mulyaningsih & Wahyu Rachmawati (eds)
 ISBN 978-0-367-33912-8

Service quality, satisfaction, and trust of patients' families in the hospital

Darti
Darti Publisher, Makassar, Indonesia

ABSTRACT: Quality service provided by a hospital will help to create feelings of satisfaction in patients and their families. The quality of hospital services can be improved through the physical environment. The purpose of this study was to analyze and assess service, comfort, satisfaction, trust, and word of mouth among patients' families at the regional special hospital of South Sulawesi Province. This study used the SEM method to examine 5–10 observations for each indicator.

Keywords: service quality, hospital image, satisfaction, trust, word of mouth

1 INTRODUCTION

Quality, according to Kotler and Keller (2012), is part of the overall characteristics of a product or service with the ability to satisfy a need. Kotler and Keller (2007) define *satisfy* as a person's level of gratification after evaluating the results of perceived performance.

Colquitt et al. (2011) define *trust* as a desire to depend on an authority based on positive expectations and the authority's attention to various aspects of life, such as competition in the world of health care facilities. Quality service is necessary to build the satisfaction and trust of patients and their families in a hospital. This helps to create feelings of satisfaction in patients or their visiting family members. Therefore, hospitals tend to compete to provide the best service to their customers.

Chang, Chen, and Lan (2013) explain that the service quality of hospitals influences the trust of patients and their families. When patients and their families consider the service to be of good quality, trust is built, and business tends to grow stronger. Alrubaiee and Alkaa'ida (2011) also show that service quality has a positive value of trust. However, based on the background of the study, the results on service quality and its impact on satisfaction and trust have not been consistent. Therefore, it is essential to examine service quality and its impact on patients and their families.

2 QUALITY OF SERVICE, SATISFACTION, AND PATIENT AND FAMILY TRUST

2.1 *Quality of service*

2.1.1 *Definition of service quality*

According to Parasuraman, Zeithaml, and Berry (1988) and Tjiptono, Chandra, and Adriana (2008), *service quality* is when a company meets customer needs and expectations.

2.1.2 *Health services*

Health comes from the word *healthy*, and humans are said to be healthy when in perfect condition with no weaknesses or defects, both physically and spiritually (Notoatmodjo, 2005).

Azwar (2008) defines *health services* as individual or joint efforts within an organization to maintain, improve, prevent, and treat diseases of individuals, groups, and the community. Therefore, health services are efforts carried out both individually and jointly by institutions to adequately manage the public.

2.1.3 *Quality of health services*

Good quality is related to recovery from illness, degree of health, the flow of services, satisfaction with health care facilities, and acceptable rates. Patients tend to value the quality of service using different standards or criteria.

2.1.4 *The dimension of service quality*

According to Zeithaml, Bitner, and Gremler (2000), 10 main factors, or SERVQUAL dimensions, determine service quality: tangibility, reliability, responsiveness, competency, credibility, courtesy of staff to customers, security (which makes customers feel safe), access, communication, and understanding the customer. These 10 factors are further divided into 5 SERVQUAL dimensions by Parasuraman et al. (1988): tangibility, reliability, responsiveness, assurance, and empathy.

2.2 *Patient family satisfaction*

2.2.1 *Definition of satisfaction*

According to Kotler and Keller (2007), *satisfaction* is an individual's level of gratification after comparing performance or perceived results to expectations.

2.2.2 *Satisfaction benefits*

Tjiptono et al. (2008) state that the creation of consumer satisfaction provides several benefits, as follows: the relationship between consumers and companies becomes harmonious; there is a good basis for repurchase intention and consumer loyalty; and word-of-mouth reputations are established.

2.2.3 *Factors affecting satisfaction*

Budiastuti (2002) suggests that patients, when evaluating products or services received, refer to several factors, including quality of a product or service, emotion, and price. Meanwhile, other experts such as Moison, Walter, and White (Haryanti, 2000) mention factors influencing consumer satisfaction, namely: product characteristics, price, service, location, facilities, image, visual design, atmosphere, and communication.

2.2.4 *Measuring satisfaction levels*

According to Kotler and Keller (2007), researchers can use various methods to measure customer satisfaction: (1) Provide a feedback/complaint system (forms, suggestion boxes, comment boxes, etc.) through which customers can communicate positive input or grievances about products or services. (2) Conduct customer surveys. (3) Hire ghost shoppers to pose as customers in order to better evaluate the experience of shopping for and using a product or service. (4) Perform an analysis of former customers who are no longer using a service or product in an attempt to learn why they left and correct the issues.

2.3 *Patient family trust*

2.3.1 *Definition of trust*

According to McShane and Glinow (2008), *trust* refers to a person's positive expectation of others in a place. It also means talking about the fate of someone or another group.

2.3.2 *Types of trust*

Sako (1997: 3) divides belief groups into three categories: (1) Contractual trust: Will the other party honor their agreement? (2) Trust in competency: Is the other party able to carry out the things it promises? (3) Trust in goodwill: Does other party take the initiative in the commitment and not indulge in harmful things?

2.3.3 *Elements of trust*

According to McKnight and Chervany, as cited by Salo and Karjaluoto (2007), trust has six main elements: beliefs, intention, behavior, system, disposition, and situational decisions.

2.3.4 *Factors affecting trust*

Hurley (2011) opines that people make decisions to trust by understanding the mental calculations behind it and choosing whether to believe or disbelieve. The three factors referred are: (a) risk tolerance – a person's level of risk tolerance has an impact on his or her willingness to trust; (b) level of adjustment – people feel comfortable with themselves and the world; (c) relative power – assuming the trustee is in a position of authority, they tend to be able to sanction those betraying their trust. However, when there is no authority, then the trustee is in a vulnerable position, thereby making it difficult to trust.

2.3.5 *Ways to increase trust*

Trust is developed with three aspects – namely *integrity*, which consists of honesty and truth; benevolence; and propensity. Integrity also means having consistency between words and conduct. *Benevolence* means people can be trusted to have an interest at heart, shown through acts of care and support. *Propensity* is related to trust, and to the extent to which individual employees trust their leaders (Robbins and Judge, 2007).

2.3.6 *Trust measurement indicator*

Mayer et al. and Rindings et al., as quoted by Casaló, Flavián and Guinalíu (2007), state that trust consists of three aspects. *Competency* relates to consumer perceptions of the skills and knowledge of others (Coulter and Coulter, in Casaló et al., 2007). *Honesty* is the belief that the second party will keep his word and fulfill it with sincerity (Gundlach and Murphy, in Casaló et al., 2007). *Virtue* means that someone is interested in the welfare of others, or that another party is willing to make an effort to achieve a common goal.

3 CONCLUSION

Based on the results obtained from the previous literature reviews, better service quality was provided, with higher satisfaction and greater trust felt by patients and their families at the regional special Hospital of South Sulawesi Province.

REFERENCES

Alrubaiee, L. and Alkaa'ida, F. (2011). The mediating effect of patient satisfaction in patients' perceptions of the healthcare quality–patient trust relationship. *International Journal of Marketing Studies*, 3(1), p. 103.

Azwar, S. (2008). *Metode Penelitian Sosial*. Yogyakarta: Pustaka Belajar.

Budiastuti (2002). *Kepuasan Pasien Terhadap Pelayanan Rumah Sakit*. Jakarta: Trans Info Media.

Casaló, L., Flavián, C., and Guinalíu, M. (2007). The impact of participation in virtual brand communities on consumer trust and loyalty: The case of free software. *Online Information Review*, 31(6), pp. 775–792.

Chang, C.-S., Chen, S.-Y., and Lan, Y.-T. (2013). Service quality, trust, and patient satisfaction in interpersonal-based medical service encounters. *BMC Health Services Research*, 13(1), p. 22.

Colquitt, J. et al. (2011). *Organizational Behavior: Improving Performance and Commitment in the Workplace*. New York: McGraw-Hill Irwin.

Haryanti (2000). *Manajemen Mutu Pelayanan Kesehatan*. Surabaya: Universitas Airlangga.

Hurley, R. F. (2011). *The Decision to Trust: How Leaders Create High-Trust Organizations*. Hoboken, NJ: Wiley.
Kotler, P. and Keller, K. (2012). *Manajemen Pemasaran. Edisi 12*. Jakarta: PT. Indeks.
Kotler, P. and Keller, K. (2007). *Manajemen Pemasaran Edisi 12 Jilid 1*. Jakarta: Indeks.
McShane, S. L. and Glinow, M. A. V. (2008). *Organizational Behavior*. New York: McGraw-Hill.
Notoatmodjo, S. (2005). *Promosi Kesehatan Teori dan Aplikasi*. Jakarta: Rineka Cipta, pp. 52–54.
Parasuraman, A., Zeithaml, V., and Berry, L. L. (1988). Konsep dan teknik pengukuran kualitas produk jasa. *Kajian Bisnis dan Manajemen*, 4(1), pp. 55–56.
Robbins, S. P. and Judge, T. A. (2007). *Perilaku Organisasi*. Jakarta: Salemba Empat.
Salo, J. and Karjaluoto, H. (2007). A conceptual model of trust in the online environment. *Online Information Review*, 31(5), pp. 604–621.
Tjiptono, F., Chandra, G., and Adriana, D. (2008). *Pemasaran Strategik*. Yogyakarta: CV Andi Offset.
Zeithaml, V., Bitner and Gremler (2000). *Service Marketing*. Singapure: Mc Graw-Hill.

Facing Global Digital Revolution – Nirmala Arum Janie,
Dwi Mulyaningsih & Wahyu Rachmawati (eds)
 ISBN 978-0-367-33912-8

Impact of posting purchases on social media behavior: Roles of materialism and purchase types

Alifa Nisfiyani
Universitas Indonesia, Depok, Indonesia

ABSTRACT: This research aimed to describe the impact of posting purchases on social media behavior and how this behavior can affect consumer happiness. The study consisted of quantitative research with a casual design. The analysis was based on an online survey of 259 respondents who had posted their purchases to social media in the previous 30 days. The result of this research showed that materialism significantly affects the tendency to post purchases. The types of purchases do not considerably change because of this behavior, however. Based on the research, the posting behavior is not significantly affected by consumers' happiness with their purchases.

Keywords: posting purchase, purchase type, materialism, consumer happiness

1 INTRODUCTION

The rapid growth of digital media in recent years has happened in almost all countries in the world, including Indonesia. Based on data from the Global Digital Report 2018, the total number of Internet users in Indonesia has reached 132 million people. Almost 50% of Indonesian citizens have access to the Internet, mostly via their smartphones (GNFI, 2018).

Research from We Are Social and Hootsuite has shown that Indonesia ranks third in the world in time spent on social media. On average, Indonesia's citizens spend more than three hours per day on social media platforms. The platforms that have seen rapid growth in users during the past years include Facebook and Instagram (Global Digital Report, 2018).

Social media users carry out many activities; one of them is sharing and creating user-generated content (Hautz, Füller, Hutter, and Thürridl, 2014). Posting purchases on social media provides an opportunity for consumers to express feelings regarding their purchases to friends. It is also part of word of mouth (WOM) and conspicuous consumption (Duan and Dholakia, 2018).

Consumption-related posting behavior (CPB) is also influenced by the type of purchase, which can be divided into two: material purchases and experiential purchases. Besides the type of purchase (based on Richins and Dawson, 1992), the level of materialism also can influence CPB.

This article seeks to investigate what determines CPB and how this behavior influences consumer happiness as well as interpersonal relationships.

2 LITERATURE REVIEW

2.1 *Purchase type*

Based on consumer purchase intention, we can categorize purchases into two types, which are material purchases and experiential purchases. Consumers make material purchases with the primary intention of acquiring material things, and they select experiential purchases with the main goal of acquiring a life experience, an event, or a series of events (Van Boven and Gilovich, 2003).

Carter and Gilovich (2012) state that the most significant difference between material and experiential purchases is consumers' intention behind them. Material purchases focus on possessions and ownership while experiential purchases focus on processes and experiences. From an economic perspective, material purchases are related to the manufacturing industry and experiential purchases are related to the service industry (Peng and Ye, 2015).

2.2 *Materialism*

We define *materialism* as the tendency to view acquisitions as the necessary means to reach important life goals and desired end states (Richins and Dawson, 1992). Consumers with higher levels of materialism are more likely to use material objects they own as a sign of their success, desire to own possessions that impress others, and admire people who own expensive possessions (Duan and Dholakia, 2018).

Consumers who have lower levels of materialism tend to care about self-expression, happiness, and doing something only for themselves. Unlike more materialistic consumers, less materialistic consumers do not consider the acquisition of goods as a path to personal happiness (Richins and Dawson, 1992).

2.3 *Posting purchase behavior*

Posting purchases on social media is one of the new types of electronic WOM behavior, which we define as oral, informal, person-to-person communication between a perceived noncommercial storyteller and a receiver regarding a brand, a product, an organization, or a service. Word of mouth has been researched extensively in terms of its influences on receivers and its effects on storytellers (Duan and Dholakia, 2017).

Posting purchases on social media is also a digital form of conspicuous consumption. *Conspicuous consumption* is showing off wealth in order to maintain social status through publicly consumed goods (Trigg, 2016). As a digital vehicle for WOM and conspicuous consumption, posting purchases on social media has characteristics that differentiate it from traditional WOM and conspicuous consumption (Duan and Dholakia, 2017).

2.4 *Consumer happiness*

We categorize consumer happiness into two types. The first type is consumer happiness that appears specifically during the use of a product or consumption of an experience (Richins, 1997). The second type is consumer happiness that arises because of the positive quality of life consumers perceive as a result of a purchase (Van Boven and Gilovich, 2003).

Scholars have also researched happiness as a positive quality of life that is influenced by purchases; purchasing more effectively helps build a person's identity and also develops social relationships with people around consumption (Van Boven and Gilovich, 2003).

3 RESEARCH METHODOLOGY

Researchers used a snowball sampling technique in this study. Respondents had posted their purchases on social media within the previous 30 days. Moreover, they were between 18 and 55 years old. We distributed the questionnaires to the selected respondents through social media, and the respondents filled it out by themselves. We focused this survey on respondents who live around Jabodetabek (the greater Jakarta area).

Hypotheses about materialism, CPB, and consumers' happiness with purchases were tested using a structural equation model (SEM) to see the impact of each variable. The illustrated model of this research is as follows:

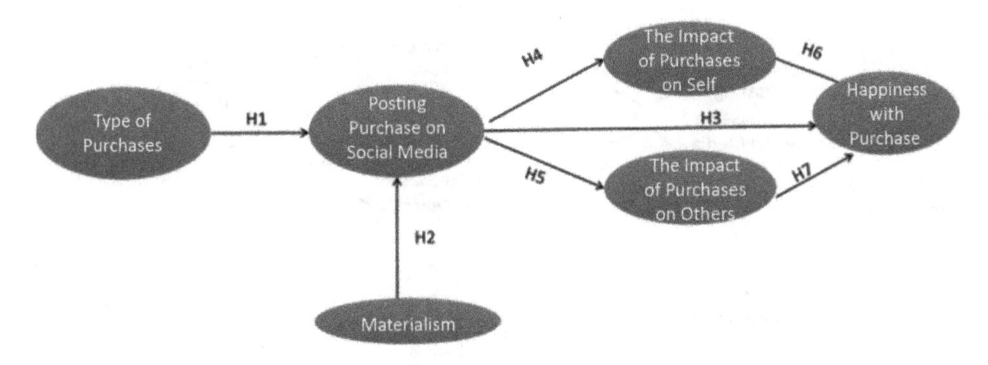

Figure 1. Research model.

4 RESULTS

This research included around 259 respondents with the same proportion between male and female. After collecting the results, a pretest was run using the Lisrel application. The result of the pretest showed that all variables were valid and reliable to use.

Table 1. Results of pretest.

Variable	Numbers	Cronbach's Alpha	Notes
Happiness with Purchase	3	0.738	Reliable
Impact on Self	6	0.906	Reliable
Impact on Intrapersonal Relationships	3	0.853	Reliable
Materialism	10	0.861	Reliable
CSWOM Tendency	6	0.883	Reliable

From testing the hypotheses, we found that of six hypotheses in this research, we could accept only four. The accepted hypotheses were H1, H3, H4, H5, while the hypotheses not accepted were H2 and H6.

The data supported H1, which states that materialism influences CPB. The data also supported H3, which states that CPB could influence the impact of purchases on consumers themselves.

The data did not support H2. That means, according to this research, that CPB did not influence consumers' happiness with their purchases. The data also did not support H6, which states that the impact of purchases on others could influence consumers' happiness with their purchases.

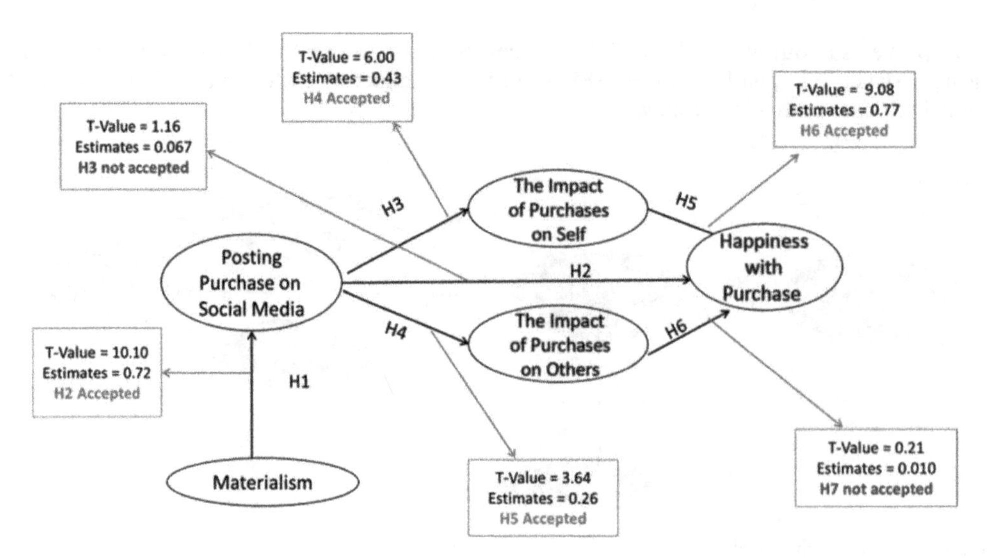

Figure 2. Structural model.

5 CONCLUSION

Materialism influences CPB, but this behavior does not always mean that a consumer is happy with a purchase. Based on the research, consumers' happiness with a purchase is more heavily influenced by the impact of purchases on the consumers themselves than by the impact of purchases on their interpersonal relationships.

Managerially, this research suggests that marketers should encourage consumers to post their purchases on social media and get rewards for that, an engagement that is useful for creating brand loyalty.

REFERENCES

Duan, J. and Dholakia, R. R. (2017). Posting purchases on social media increases happiness: The mediating roles of purchases' impact on the self and interpersonal relationships. *Journal of Consumer Marketing*, 34(5), pp. 19–39.

Duan, J. and Dholakia, R. R. (2018). How purchase type influences consumption-related posting behavior on social media: The moderating role of materialism. *Journal of Internet Commerce*, 17(1), pp. 64–80.

Hautz, J., Füller, J., Hutter, K., and Thürridl, C. (2014). Let users generate your video ads? The impact of video source and quality on consumers' perceptions and intended behaviors. *Journal of Interactive Marketing*, 28(1), pp. 1–15.

Peng, Z. and Ye, M. (2015). An introduction of purchase types and happiness. *Journal of Service Science and Management*, 8(1), pp. 132–141.

Richins, M. L. and Dawson, S. (1992). A consumer values orientation for materialism and its measurement: Scale development and validation. *Journal of Consumer Research*, 19(3), p. 303.

Trigg, A. B. (2016). *Veblen, Bourdieu, and Conspicuous Consumption.*

Van Boven, L. and Gilovich, T. (2003). To do or to have? That is the question. *Journal of Personality and Social Psychology*, 85(6), pp. 1193–1202.

Facing Global Digital Revolution – Nirmala Arum Janie, Dwi Mulyaningsih & Wahyu Rachmawati (eds)
 ISBN 978-0-367-33912-8

Predator and prey: Ponzi and pyramid investors

Taofik Hidajat
Sekolah Tinggi Ilmu Ekonomi Bank BPD Jawa Tengah, Semarang, Indonesia

ABSTRACT: This paper proposes to present a new perspective on the tricky investment practices through the classification of investors' Ponzi and pyramids scheme. This study used qualitative methods through interviews with victims of the scheme from 11 cities in Indonesia. The sample in this study were individuals who were victims of investment fraud with both systems. The method applied was snowball sampling. The small number of samples was a limitation of the study. Based on the interviews, there were three classifications of investors of both schemes: Predator, Strong Prey and Weak Prey. The current study might be expected to be the first research on investor classification on Ponzi and pyramid scheme.

1 INTRODUCTION

Illegal investment "getting rich quick scheme" through Ponzi and the pyramid schemes still exist. The victims of this scheme can be found not only in developing countries but also in developed countries. There were more than 500 bad impacts of Ponzi schemes during 2008-2013 in The United States (Maglich, 2014) and 700 illegal businesses practices during June 2007 that were dissolved by the Malaysian government (Sulaiman, Moideen and Moreira, 2016). The same schemes also exist in Russia, Bulgaria, Romania, Serbia, Slovakia, Czech Republic, Albania (Bhattacharya, 2003), China (Albrecht et al., 2017), Indonesia (Hidajat, 2018), et cetera.

In Indonesia, some companies that operate Ponzi and pyramid schemes were part of 262 illegal investments as released by the Indonesian Financial Services Authority in 2014. From 262 illegal companies, 52 of them were operating Ponzi and pyramid schemes. This amount is too little because there are many companies that are not detected. When it first appeared in Indonesia, the most widely used scheme was the pyramid through chain letters. Although there is no official record of the presence of this scheme in Indonesia, the presence of Bratajaya Nilawati in 1975 showed that this mode of investment had emerged that year. The method used at that time was through chain letters whose payments were made through money orders.

Along with people's changing behavior, preferring to get results quickly, the popularity of the pyramid scheme is fading and has been replaced by Ponzi schemes. Instruments that come in the guise of investment, in general, are forex, gold commodities or other derivative transactions. On the internet, this scheme often arises through high yield investment programs (HYIP).

Some studies had been conducted on the existence of Ponzi and pyramid schemes in several countries, such as Caribbean (Carvajal et al., 2009), Jamaica (Tennant, 2011), South Africa (Krige, 2012), United States (Deason, 2012), and Papua New Guinean (Cox, 2014). There is also research that discusses the high yield investment program (HYIP) - a Ponzi and Pyramid scheme using the internet - like Moore, Han, & Clayton (2012) and Artzrouni (2009). Other research addresses efforts to avoid schemes (Baucus and Mitteness, 2016) through regulation (Sulaiman, Moideen and Moreira, 2016) and the causes of investors joining the scheme (Jacobs and Schain, 2011; Tennant, 2011; Lewis, 2012; Wilkins, Acuff and Hermanson, 2012).

However, to our knowledge, there has been no research that discusses the classification of investors in Ponzi and pyramid schemes. This classification is important because not all investors in both schemes have the same knowledge and motivation. For example, Moore, Han, & Clayton (2012) stated that HYIP is 'postmodern investment' because investors who participate in the program know the fraud committed by the schemer, but they still take risks by joining early and taking profit at the right time. This paper aims to classify Ponzi and pyramids investors. This classification is important because it is useful to know the characteristics of each group.

2 PONZI AND PYRAMID SCHEMES

Innovations in financial markets did not only give rise to various types of trust investment, but they were also able to create negative creativity with the existence of illegal investment schemes. The phenomenon of "getting rich quickly" through Ponzi and pyramid schemes was one of the illicit investments that have been occurring continuously until now.

A Ponzi scheme is "a type of investment fraud in which returns are paid to investors either from their own money or out of money paid in by subsequent investors, rather than from profits generated by investment or any genuine business activity" (Lewis, 2012). In a Ponzi scheme, the schemer will search potential investors to join. Investors only invest their money and wait for profit sharing without doing anything. According to Parsons (2011), some characteristics of the classic Ponzi scheme are: (i) The source of income for this scheme comes only from investors; (ii) exclusivity. To give the impression that the system is exclusive and not easy to follow, only certain investors are allowed to follow the scheme; (iii) Affinity fraud. Schemer will generally look for targets, namely new investors from specific communities based on religion, race, age, and so on. In Madoff's case, many investors came from the Jewish community and the stock investor community; (iv) Consistency. The scheme consistently fulfils the promise of providing investment returns to old investors and attracting new investors; (v) Secret element. The schemer will never explain how to get income consistently. In general, they will claim to have several investment strategies; (vi) Perceived legitimacy. Schemer reputation factors and have been operating for a long time, making investors feel confident that their funds are in the right hands.

Although the profit received by investors comes from other investors, the schemer usually generally states that the return on investment is a result of legal investment (Deason, 2012) or the result of applying a specific investment strategy by the schemer. Bernard Madoff for example, publicly stated that he used a split-strike investment strategy in managing investor funds (Bernard, 2009; Clauss, Roncalli and Weisang, 2009; Gregoriou, 2009). The schemer usually also uses trust factors such as public figures to attract attention. Prospective investors who have confidence in certain individuals tend not to carry out investigations to detect fraud (Carey and Webb, 2017).

The pyramid scheme is similar to a Ponzi scheme where returns received by an investor also come from other investors money. According to Akinladejo et al. (2013), "A pyramid scheme is often described as a 'business opportunity'. The distinctive characteristic of this 'business opportunity' is that the only way participants can make money is by recruiting other members to the scheme who quickly find out that their successes depend entirely on their ability to recruit other persons to the scheme." In contrast to the Ponzi scheme where the managers act actively to find investors, in the pyramid scheme, investors are the parties that actively seek another investor. If new investors are not obtained, they get nothing.

In a pyramid scheme, there are two types of models: the naked pyramid and the product-based pyramid (Akinladejo, Clarke and Akinladejo, 2013). In the naked pyramid, the existing mechanism is derived from the recruitment of new investors. To give the good image that this investment scheme is true or logical, the product-based pyramid model incorporates the product as camouflage. For the ordinary people with low level of financial knowledge, this model is sometimes complicated to distinguish from legal and true investment because there are other products or services provided, so it looks as multi-level marketing (MLM) business.

From the discussion, the Ponzi and pyramid schemes can be defined as "investment schemes that promise high or certain returns level to investors, but really, the rewards or returns given to those investors were derived from other investors money".

3 METHOD

This research uses interviews through questionnaires with victims of Ponzi and pyramids schemes. The sample in this study were those who were or had participated in one or more investment offers from high yield investment programs (HYIP) and 52 companies that run Ponzi and pyramid schemes. The fifty-two companies are part of a list of 262 companies considered illegal by the Indonesian Financial Services Authority in 2014.

The sampling technique used is snowball sampling because it is difficult to find victims of investment fraud as stated by Sadiraj & Schram (1999). We found respondents one by one respondent based on information from other victims of the scheme. The number of samples obtained was 98 people from 11 cities in Indonesia consisting of 17% of the pyramid scheme investors and 83% of the Ponzi scheme.

4 RESULTS AND DISCUSSIONS

The interview results illustrate that the average age of pyramid scheme investors is 50. This is an indication that this scheme has been around for a long time being the most numerous compared to other schemes. Almost all the respondents in this category are pensionary and risk avoiders. This finding is consistent with Bosley & Knorr (2017) that those who have just retired are one of the groups targeted by the schemer. They want to get money but do not want to lose money from an investment.

Unlike pyramid investors, almost all Ponzi (non-HYIP) scheme investors are adults with an average age of 45 years and Ponzi investors (HYIP) are young people with an average age of 24 years. This result is in line with the existence of the scheme in Indonesia which in recent years has been dominated by Ponzi schemes. Ponzi (non-HYIP) scheme investors tend to be neutral to risk while Ponzi investors (HYIP) are risk seekers. It was surprising that 85% respondents did not know the investment they participated in included a Ponzi or pyramid scheme, while 15% of the respondents who were all Ponzi investors (HYIP) knew about this. From the description of the respondent, the characteristics of the Ponzi and pyramid scheme investors are (i) almost all investors in pyramid schemes are old, risk avoiders and do not know the mechanism of fraud committed by the schemer; (ii) Ponzi investors consist of two groups, Ponzi HYIP and Ponzi non-HYIP. Ponzi HYIP consists of young people and risk takers because they know how the schemer works. They are difficult to influence and join investments because of their wants.

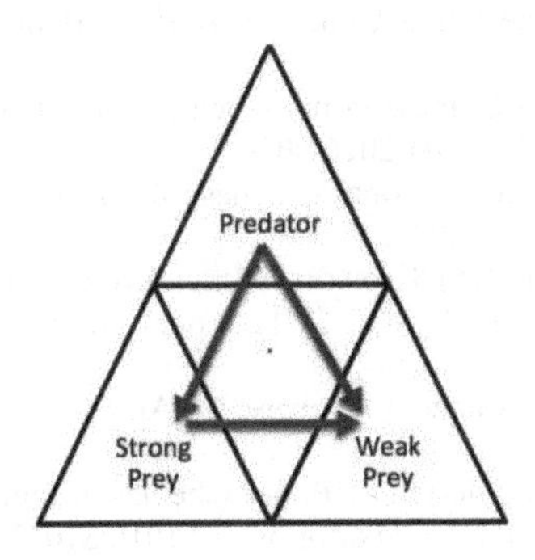

Figure 1. Ponzi and pyramid investor.

From these results, this paper proposes three types of investors in the Ponzi and pyramid schemes, namely Predators, Strong Prey and Weak Prey. The terms are based on John Maynard Keynes's statement that "animal spirit" could be implemented to describe human behaviour such as bullish, bearish, herding, et cetera. Predator is a term for investors who understand fraudulent schemes but deliberately take risks by joining earlier to get more profits. Predators know that the system is a very risky business scheme. Predators exist in both Ponzi and pyramid schemes. They not only want to find new businesses that appear with Ponzi schemes or pyramids but also actively seek other investors (prey) to join, especially in pyramid schemes. Most of them have experience in the same scheme. They also understand when they must enter (participate) and exit (profit taking). They are often sought out by schemers of pyramid schemes to become leaders that have just emerged and to influence others (prey) to join.

Strong Prey was a term attached to investors who did not understand the scheme mechanism, but they joined in Ponzi or pyramid scheme more due to their intention. Strong Prey were in Ponzi and pyramid schemes. They considered that the business scheme was followed was a legal business or investment. Although they tended to be Prey for Predators, Strong Prey was difficult to influence. Weak Prey was a term for investors who did not understand the scheme mechanism, but they joined in it because of the persuasion, influence or just following the others. Generally, Weak Prey were present in the pyramid scheme, and they were at the latest line in the system. They were the parties who were almost always the victims of this tricky scheme because of their ignorance, and others very easily influenced them. They were also easily moved from one scheme to another because of persuasion from others. The existence of weak prey indicated the presence of "unexpected hanging paradox". It was a condition wherein a pyramid scheme, almost certainly no one wants to join the latest, but in real fact, there were always parties who decided to join.

5 CONCLUSIONS

Illegal investment offers through Ponzi and pyramid schemes continue to occur using several tricks. The existence of these schemes cannot be separated from investor profiles in each scheme. Learning from cases in Indonesia, Ponzi investors or pyramid schemes are grouped into three types: Predator, Strong Prey, and Weak Prey.

Predators are the type of investor who best understands the scheme and influences Strong and Weak Prey to join. Even so, Strong Prey is prey that is more difficult to influence than Weak Prey. Weak Prey is a group of investors who get nothing from a scheme.

REFERENCES

Akinladejo, O. H., Clarke, M. and Akinladejo, F. O. (2013) 'Pyramid Schemes and Multilevel Marketing (MLM): Two Sides of the Same Coin', *Journal of Modern Accounting and Auditing*, 9(5), pp. 690–696.

Albrecht, C. et al. (2017) 'Ezubao: a Chinese Ponzi scheme with a twist', *Journal of Financial Crime*, 24 (2), pp. 256–259. doi: doi:10.1108/JFC-04-2016-0026.

Artzrouni, M. (2009) 'The mathematics of Ponzi schemes', *Mathematical Social Sciences*, 58(2), pp. 190–201. doi: 10.1016/j.mathsocsci.2009.05.003.

Baucus, M. S. and Mitteness, C. R. (2016) 'Crowdfrauding: Avoiding Ponzi entrepreneurs when investing in new ventures', *Business Horizons*, 59(1), pp. 37–50. doi: http://dx.doi.org/10.1016/j.bushor.2015.08.003.

Bernard, C. (2009) 'Mr . Madoff ' s Amazing Returns: An Analysis of the Split-Strike Conversion Strategy', pp. 1–30.

Bhattacharya, U. (2003) 'The optimal design of Ponzi schemes in finite economies', *Journal of Financial Intermediation*, 12(1), pp. 2–24. doi: http://dx.doi.org/10.1016/S1042-9573(02)00007-4.

Bosley, S. and Knorr, M. (2017) 'Pyramids, Ponzis and fraud prevention: lessons from a case study', *Journal of Financial Crime. Emerald*, 25(1), pp. 81–94. doi: 10.1108/JFC-10-2016-0062.

Carey, C. and Webb, J. K. (2017) 'Ponzi schemes and the roles of trust creation and maintenance', *Journal of Financial Crime. Emerald*, 24(4), pp. 589–600. doi: 10.1108/JFC-06-2016-0042.
Carvajal, A. et al. (2009) 'Ponzi Schemes in the Caribbean.
Clauss, P., Roncalli, T. and Weisang, G. (2009) 'Risk management lessons from Madoff Fraud', *International Finance Review*, 10, pp. 505–543.
Cox, J. (2014) 'Fast Money Schemes Are Risky Business : Gamblers and Investors in a Papua New Guinean Ponzi Scheme', 84(3), pp. 289–305. doi: 10.1002/ocea.5062.
Deason, S. E. (2012) 'Ponzi Scheme Lifecycles: An Initial Foray into the Determinants of Ponzi Scheme Fraud Lifespan and Size', *Goizueta Business School, Emory University*.
Gregoriou, G. N. (2009) 'Madoff : A Riot of Red Flags', 33(January).
Hidajat, T. (2018) 'Financial Literacy, Ponzi and Pyramid Scheme in Indonesia', *Jurnal Dinamika Manajemen*, 9(2), pp. 198–205.
Jacobs, P. and Schain, L. (2011) 'The Never Ending Attraction of the Ponzi Scheme'.
Krige, D. (2012) 'Fields of dreams, fields of schemes: ponzi finance and multi-level marketing in South Africa', *Africa*, 82(01), pp. 69–92. doi: 10.1017/S0001972011000738.
Lewis, M. K. (2012) 'New dogs, old tricks . Why do Ponzi schemes succeed ? ☆', *Accounting Forum. Elsevier Ltd*, 36(4), pp. 294–309. doi: 10.1016/j.accfor.2011.11.002.
Maglich, J. (2014) 'A Ponzi Pandemic: 500+ Ponzi Schemes Totaling $50+ Billion in 'Madoff Era'. Available at: http://www.forbes.com/sites/jordanmaglich/2014/02/12/a-ponzi-pandemic-500-ponzi-schemes-totaling-50-billion-in-madoff-era/#74099f3c10b9.
Moore, T., Han, J. and Clayton, R. (2012) 'The postmodern Ponzi scheme: Empirical analysis of high-yield investment programs', *International Conference on Financial Cryptography and Data Security. Springer*, pp. 41–56.
Parsons, L. (2011) 'Gullible, Greedy or Just Unlucky? How Bernie Madoff Scammed About 15,000 Investors', The National Legal Eagle, 16(1), p. 3. Available at: http://epublications.bond.edu.au/nle/vol16/iss1/3.
Sadiraj, K. and Schram, A. (1999) 'Informed and uninformed investors in an experimental ponzi scheme. University of Amsterdam', *Working Paper*.
Sulaiman, A. N. ., Moideen, A. I. and Moreira, S. D. (2016) 'Of Ponzi schemes and investment scams A case study of enforcement actions in Malaysia.', *Journal of Financial Crime*., 23(1), pp. 231–243. doi: 10.1108/JFC-05-2014-0021.
Tennant, D. (2011) 'Why do people risk exposure to Ponzi schemes? Econometric evidence from Jamaica', *Journal of International Financial Markets, Institutions and Money*, 21(3), pp. 328–346. doi: http://dx.doi.org/10.1016/j.intfin.2010.11.003.
Wilkins, A. M., Acuff, W. W. and Hermanson, D. R. (2012) 'Understanding a Ponzi Scheme: Victims ' Perspectives', 4(1), pp. 1–19.

Facing Global Digital Revolution – Nirmala Arum Janie, Dwi Mulyaningsih & Wahyu Rachmawati (eds)
© 2020 Taylor & Francis Group, London, ISBN 978-0-367-33912-8

Effect of transparency and accountability on employee performance in OPD in Semarang city using budgeting based performance as an intervening variables

A. Vivianita, D. Indudewi & F.N Prihantini
Universitas Semarang, Semarang, Indonesia

ABSTRACT: Cases of deviant budgeting occur in many regions in Indonesia. The case of budgeting is not based on performance but it is arranged carelessly. The original arrangement caused poor budget absorption so that the realization and absorption of the budget for activities, development, and service to the community were not optimal. In addition, there is poor governance when preparing budgets and realizing the budget. The purpose of this study was to analyze the effect of budgeting based performance, transparency and accountability on the performance of the OPD, and to analyze the effect of transparency and accountability on performance of the OPD through budgeting based performance. This research uses the contingency theory. The sample in this study used the accounting or financial department working in the Regional Government Organization (OPD) of the Semarang city. Sampling uses the random sampling method. Research method uses Structural Equation Modeling (SEM). The statistical tool is Smart PLS 3.0. The results of the study are that transparency has a significant effect on OPD performance, accountability does not affect the performance of the OPD, performance-based budgets affect the performance of the OPD, transparency does not significantly influence OPD performance through budgeting based performance, and accountability influences performance significantly through budgeting based performance.

Keywords: transparency, accountability, budgeting based performance and performance of OPD

1 INTRODUCTION

The cases of budget processes that are not transparent and closed cause the performance of OPD to decline in the eyes of the public. The corruption and budget leakage in Maluku often occur because budgeting is not carried out transparently and accountably, but in closed spaces, so that budget haggling occurs. Soedasono Tandra who is chairman of the Association of Indonesian Curators and Administrators, emphasized that budget leakage and budget corruption in Maluku will not occur when the budget discussion is conducted more openly or transparently and accountably, so that the public can control the regional government budget (rakyatmaluku.com, 2019).

The budgeting process that is not transparent and careless, and is not based on the performance and participation of related parties, causes the realization and uptake of the budget cannot be used maximally to the public needs. Some cases of budget absorption that are not as optimal as the budgeting process happened in Jakarta. In 2016, DKI Jakarta APBD budget are not absorbed for almost Rp. 8 trillion so that development and community services cannot be conveyed properly (news.detik.com.2016). Besides DKI Jakarta, Semarang city is also unable to absorb and realize a well-prepared budget. This causes a loss for the community. Tribune Jateng.com (2017) reported that the absorption of the Semarang City Government budget in the middle of 2017 was only 27%, in fact, the city should have absorbed a budget of 50%. The Chairperson of the Semarang City DPRD (Supriyadi) also confirmed that from the total regional budget of Rp. 4.5 trillion in the middle of 2017, but only Rp. 1.2 trillion was realized (Permadi. Tribun Jateng.com, 2017). Some cases that occur due to the budgeting process that is covered for group interests are not based on participatory and performance.

Performance based budgeting is the allocation of funds used in accordance with the objectives of the organization's activities, vision, mission and goals that are carried out effectively and efficiently (Young, 2003 in Ibrahim, 2015). However, the practice of this performance based budgeting is not optimal because there are still cases such as the presence of budget brokers (Fauzia, www.kompas.com, 2018) and corrupt practices when preparing budgets, as well as evident from budget absorption that is not optimal for public services and development (Tribun Jateng.com, 2017). In addition to several cases that occurred, there were some inconsistencies in the results of the study.

2 RESEARCH FRAMEWORK

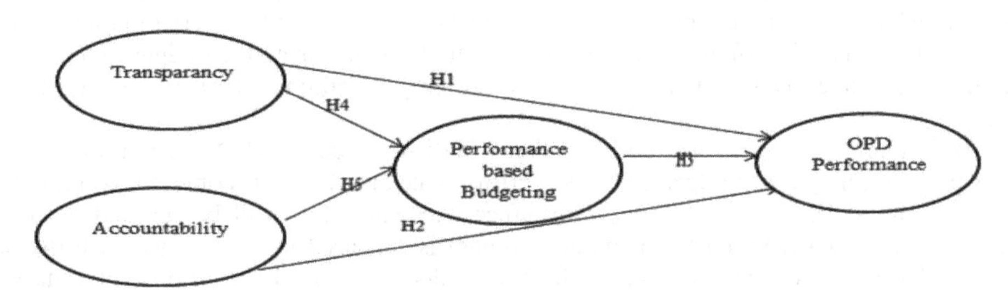

Based: development from research in 2018

2.1 *Research methodology*

Quantitative research used in the research. The samples is treasurer or financial section in the OPD in the city of Semarang. Data is collected using a questionnaire. The analytical method is Structural Equation Modeling. The statistical tool using Smart PLS 3.0.

2.2 *Research results*

2.2.1 *Outer model*

Testing the outer model are validity and reliability. The validity test uses of the AVE (Average Variant Extract) test and reliability uses the test Cronbach alpha. The test results show all indicators of each variable have a value of > 0.5, which means the research indicators are valid and reliable.

2.2.2 *Inner model*

R Square

The test results show that the value of R Square of budgeting based performance is 0.842 or equal to 84.2%. While the performance is 0.791 or equal to 79.1%.

3 DISCUSSION OF RESEARCH RESULTS

Hypothesis 1 has a T value of 2.034> 1.96 and P-Value 0.042 <0.05, meaning that transparency has a significant effect on OPD performance, which means that the higher budget transparency carried out in the OPD will spur and improve the performance of the OPD. This result is supported by research from (Saputra and Darwanis, 2014; Krisherdian, R, A, 2015; Auditya and Husaini, 2013; Rohmah, 2012).

Hypothesis 2, namely accountability does not significantly affect the performance of OPD, because the value of T-Statistics is 0.102 <1.96, and P Value is 0.919> 0.05. accountability that exists at each OPD depends on the person who runs it and the person who is authorized to carry out the program. Accountability will not go well when the morale of the employees in OPD is not good, so the programs cannot be carried out according to public plans and expectations. The results of this study were supported by Adiwirya and Sudana (2015).

Hypothesis 3 shows that there is a significant influence between budgeting based performance on OPD performance because the value of T Statistics > 1.96, which is 2.904 and P-Value <0.05, which is 0.04. The performance improvement of the OPD is caused by the budget that is in line with the performance or program that will be carried out by the OPD so that the budget absorption is in accordance with the implementation schedule and there will be no programs or plans with less or more budget. The previous study of Verasvera (2016) supports this result.

The test results in hypothesis 4 shows that budgeting based performance are unable to influence the effect of transparency on OPD performance, which is seen from the value of T Statistic 0.072 <1.96 and the value of P-Value 0.626> 0.05. OPD performance increases, not because of what basis is used when preparing budgets, whether based on performance, participation or other bases, but on how transparent the OPD is when using the budget to implement programs that have been prepared for the community. Ibrahim (2015) also supports this result.

The fifth hypothesis is accepted because, the T-Statistic value is 2.698> 1.96 and P Value is 0.007 <0.05, meaning that budgeting based performance are able to become intervening variable to accountability against OPD performance. Accountability will be fulfilled when the budget that has been received from the government is prepared in accordance with the performance that will be carried out so that the budget does not have less and more, because the portion is in accordance with the program that will be realized. The implementation of the program that is in accordance with the budget that has been prepared will improve the performance of the OPD because at the time of implementation the budget is appropriate when drafting. The results of this study are supported by the research of Suhendar (2016).

4 CONCLUSION

The results showed that the performance of OPD would increase if there was high budget transparency and budgeting based performance, while accountability did not affect the performance of the OPD. In addition, budgeting based performance cannot be intervening in the relationship between transparency and OPD performance, but performance based budgeting can be intervening in the relationship between accountability to OPD performance.

REFERENCES

Adiwirya, M. F. and Sudana, I. P. (2015) 'Akuntabilitas, Transparansi, dan Anggaran Berbasis Kinerja pada Satuan Kerja Perangkat Daerah Kota Denpasar', *E-Jurnal Akuntansi*, pp. 611–628.

Auditya, L. and Husaini, L. (2013) 'Analisis pengaruh akuntabilitas dan transparansi pengelolaan keuangan daerah terhadap kinerja pemerintah daerah', *Jurnal Fairness Vol. 3*, 1.

Ibrahim, P. S. D. (2015) 'Pengaruh Anggaran Berbasis Kinerja, Transparansi, Akuntabilitas, Dan Pengawasan Internal Terhadap Kinerja Pegawai Sekretariat Jenderal DPR RI', *Jurnal Tekun*, 6, pp. 82–95.

Krisherdian, R, A, G. (2015) 'Pengaruh Transparansi dan Akuntabilitas Keuangan Daerah Terhadap Kinerja Pemerintah Kabupaten Jember', *Universitas Jember*.

News.detik.com. 2016. Cegah Main Mata di DPRD Mendagri Usul Penyusunan APBD Lewat Pergub https://news.detik.com/berita/d-4036440/cegah-main-mata-di-dprd-mendagri-usul-penyusunan-apbd-lewat-pergub (diakes tanggal 20 Agustus 2018).

Permadi. 2017. Duh Serapan Anggaran Pemkot Semarang Baru 27 Persen, Padahal Sudah Setengah Tahun. Tribun Jateng.com. https://jateng.tribunnews.com/amp/2017/08/21/duh-serapan-anggaran-pemkot-baru-27-persen-padahal-sudah-setengah-tahun.

Rohmah, L. (2012) 'Persepsi Mahasiswa Akuntansi Unesa Tentang Pengaruh Transparansi dan Akuntabilitas Keuangan Partai Politik Terhadap Kinerja Partai dan Upaya Pemberantasan Kejahatan Kerah Putih', *Universitas Surabaya.*

Rakyat Maluku.com. 2019. Tandra: Transparansi Anggaran Kunci Atasi Korupsi http://rakyatmaluku.com/2019/02/tandra-transparansi-anggaran-kunci-atasi-korupsi/(diakses tanggal 26 Fenruari 2019).

Saputra, D. S. and Darwanis, S. A. (2014) 'PENGARUH TRANSPARANSI, AKUNTABILITAS DAN KOMITMEN ORGANISASI TERHADAP KINERJA SATUAN KERJA PERANGKAT DAERAH (Studi Pada Pemda Kabupaten Aceh Selatan)', *Jurnal Administrasi Akuntansi: Program Pascasarjana Unsyiah*, 3(2).

Suhendar, D. (2016) 'Pengaruh Penerapan Prinsip-Prinsip Good Governance Terhadap Keberhasilan Penerapan Penganggaran Berbasis Kinerja APBD Kabupaten/Kota Se-Wilayah III Cirebon Dengan Komitmen Organisasi Sebagai Variabel Moderator', *Jurnal Riset Keuangan Dan Akuntansi*, 2(2).

Verasvera, F. A. (2016) 'Pengaruh Anggaran Berbasis Kinerja terhadap Kinerja Aparatur Pemerintah Daerah (Studi Kasus pada Dinas Sosial Provinsi Jawa Barat)', *Jurnal Manajemen Maranatha*, 15(2).

Facing Global Digital Revolution – Nirmala Arum Janie, Dwi Mulyaningsih & Wahyu Rachmawati (eds)
© 2020 Taylor & Francis Group, London, ISBN 978-0-367-33912-8

The effect of intellectual capital and dividend policies on a company's value with free cash flow as a moderation variable: Evidence from manufacturing companies listed on the Indonesia Stock Exchange (IDX)

Winarsih & Wahyu Budi Lestari
Sultan Agung Islamic University, Semarang, Indonesia

Dyah Nirmala A. Janie
Universitas Semarang, Semarang, Indonesia

ABSTRACT: This study aimed to analyze the effect of intellectual capital (IC) and dividend policies on company values using free cash flow (FCF) as a moderating variable. The companies studied included 132 manufacturing companies listed on the Indonesia Stock Exchange (IDX) in 2014–2016. The data used were secondary data. Researchers determined these samples using a purposive sampling technique. The data analysis comprised multiple linear regression using the EVIEWS program. The results of the study showed that IC and dividend policies do not significantly impact corporate value, while FCF has a positive and significant effect on firm value. Free cash flow as the moderating variable significantly influences IC, and dividend policies have a negative impact on firm value.

Keywords: intellectual capital, dividend policies, free cash flow, company value

1 INTRODUCTION

The development of business encourages many sectors to expand and compete globally. The value of companies that go public serves as a benchmark for investors. If the stock price is getting higher, then the value of a company will also increase, because the high stock value indicates high shareholder prosperity as well. The value of a company can be measured in several aspects: Tobin's Q Price Book Value (PBV), the price earning ratio (PER), the dividend yield ratio, the dividend payout ratio (DPR), and the market-to-book value of equity (MBVE). This study used Tobin's Q because the measurement describes how investors appreciate the company.

Many factors lead to a decrease in company value; according to Stewart (1997),a decrease in company value can occur because companies tend to focus more on hard or real assets than on intangible assets. Intangible assets that can increase the value of a company are known as *intellectual capital* (IC) (Guthrie and Petty, 2000). In Indonesia, IC developed after the advent of PSAK Number 19 about intangible fixed assets, which are identifiable nonmonetary assets that have no physical form and that are used to generate or deliver goods or services, leased to other parties, or serve administrative purposes. Therefore, companies need resources with good IC. Research conducted by Halim et. all (2016) suggests that IC positively affects company value. The higher the IC of a company, the higher that company's value.

One corporate financial function consists of dividend policies, which are policies whereby a company can establish the proportion of the profit received, then pay investors in accordance with the shares they own. Although the amount of dividends paid can provide assurance ofa company's value to investors, the company also needs to consider the funds needed for corporate development. The company's growth and dividend distribution are both necessary, but they are two different things involving different interests. The company can take a step to get funding through stock sheet offers by listing its shares on the Indonesia Stock Exchange (IDX). The dividend policies allocated to shareholders make the company's value continue to increase. Ayem and Nugroho (2016) indicate that dividend policies positively affect company value.

This research examined the influence of IC and dividend policies on company value with free cash flow (FCF) as a moderating variable. The moderation variables in this study serve to strengthen investor confidence in the performance of IC disclosures and dividend policies that will affect corporate value.

2 LITERATURE REVIEW

2.1 *Agency theory*

Agency theory exposes the behavior of corporate parties, such as managers and investors, where they have different interests. Agency theory contends that a company manager does not always act for the benefit of shareholders. If the financial statements reported to the shareholders are transparent, they can reduce agency problems. Agency problems occur when a company gives decision-making responsibility to an agent or to management. An investor can be a principal and a management agent or a manager of the company. Principals in this case are related to shareholders and bondholders, who are found in almost all companies.

3 RESEARCH METHOD

3.1 *Population and research sample*

The population in this research comprised manufacturing companies listed on the Indonesia Stock Exchange (IDX). The researchers employed purposive sampling based on a specific assessment (Indriantoro and Supomo, 2009). The criteria used were as follows:

1. Manufacturing companies listed on the Indonesia Stock Exchange in the period 2014–2016 as presented on sahamok.com. Sahamok.com is a reference used by brokers and other researchers that lists active stocks(Riyadi, 2018).
2. Manufacturing companies that published their financial statements consistently in the period 2014–2016.
3. Manufacturing companies that used the rupiah currency in the period 2014–2016.
4. Companies with complete component data, especially data related to the calculation of IC, FCF, dividend policies, and company value in the period 2014–2016.

4 RESEARCH RESULTS AND DISCUSSION

The results of this research were analyzed using the EVIEWS application program with the least square estimation method and the moderation effect model, with almost similar interaction. The verification results in this research are described as follows:

4.1 *The conceptual model of this research*

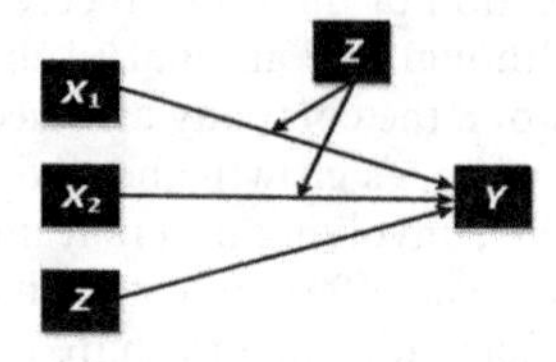

Figure 1. Research model (Source: Processed data, 2019).

Figure 1 shows groups with an exogenous or independent construct – namely IC (X_1) and dividend policies (X_2) – and a group with an endogenous construct, known as the dependent construction of the company value with a moderating variable FCF (Z).

5 DISCUSSION

Based on hypothesis testing, the researchers obtained the following results:

Table 1. Hypothesis test results.

Relation	Regression Coefficient	P-value	Summary
IC → NP	−0.009405	0.0976	Not significant for α = 5%, Significant for α = 10%
KD → NP	0.001094	0.4607	Not significant for α = 5%
FCF → NP	17.22909	0.0000	Significant for α = 1%
IC*FCF → NP	−0.130526	0.0096	Significant for α = 1%
KD*FCF → NP	−0.027542	0.0000	Significant for α = 1%

Source: Processed secondary data, 2019

5.1 *The influence of intellectual capital on company value*

Based on the research results,we can conclude that IC has no significant influence on company value. Thus the hypothesis that IC positively affects company value is rejected. These results show that a low level of IC owned by a company may not necessarily affect company value. Because of the IC measurement, no standardization can be used as a guideline; it is difficult to quantify the IC. In Indonesia, reporting IC is also voluntary (Yuskar and Novita, 2014).

This research also does not align with agency theory because the results show that the principal will not give a high price to a company based on the low level of IC. Increasing the company value is still dominated by the use of physical assets in corporate operations (Andhieka and Retnani, 2017).

5.2 *The influence of dividend policies on company value*

Based on the research results, we can conclude that dividend policies have no significant influence on company value. Thus the hypothesis that dividend policies positively affect company value is rejected. The results of this research also did not align with signal theory because the results showed that investors' response to the company is not good enough; investors will not invest in company shares based solely on the magnitude of the policies. Because dividend size is not always followed by fluctuations in corporate value, the results are synchronous under Modigliani and Miller's dividend policy theory, which proves that if dividend policies are not influenced, company value is assigned only to a company's capability to gain profit from its assets.

5.3 *The influence of free cash flow on company value*

Based on the research results, we can conclude that FCF has significant influence on company value. Thus the hypothesis that FCF positively affects corporate value is accepted. The results prove that if the FCF received by a company increases, then the value of the company also increases. Management will receive high pressure from investors to increase company value due to the FCF that companies have. As such, this certainly shows whether the company can provide prosperity for investors. The results of this research are in line with signal theory.

5.4 *Free cash flow mediates the relationship of intellectual capital to company value*

The research results showed that FCF significantly moderates the relationship between IC and company value in a negative direction. The greater the FCF, the lower the influence of IC on corporate value.

5.5 *Free cash flow mediates the relationship of dividend policies to company value*

The research results showed that FCF significantly moderates the relationship between dividend policies and corporate value in a negative direction. The larger the FCF, the lower the influence of dividend policies on company value.

6 CONCLUSIONS AND SUGGESTIONS

6.1 *Conclusions*

Intellectual capital has no significant influence on company value. Dividend policies have no significant influence on corporate value. Free cash flow has significant influence on company value. Free cash flow significantly moderates the relation between IC and company value in a positive way.

6.2 *Suggestions*

Future researchers can elaborate on this study by examining other free variables that have not been described in this research that can affect the dependent variables. Upcoming researchers are also advised to focus on sectors other than manufacturing in sampling, such as property and real estate, mining, etc.

REFERENCES

Andhieka, K. T. and Retnani, E. D. (2017). The influence of profitability and intellectual capital on company value. *Jurnal Ilmu dan Riset Akuntansi*, 6(9).

Andini, N. W. and Wirawati, N. G. P. (2014). The effects of cash flow toward financial performance and the implications for manufacturing company value. *E-Jurnal Akuntansi Universitas Udayana*, (7) 1.

Ayem, S. and Nugroho, R. (2016). The influence of profitability, intellectual structure, dividend policy and investment decisions toward company value. *Jurnal Akuntansi*,(4)1.

Brigham, E. F. and Ehrhardt, M. C. (2005). *Financial Management*. Orlando, FL:Harcourt College.

Brigham, E. F. and Houston, J. (2011). *Fundamentals of Financial Management*. Boston, MA: Cengage Learning.

Dechow, P. (1995). Accounting earnings and cash flows as a measure of firm performance: The role of accounting accruals. *Journal of Accounting and Economics*, 40, pp. 3–42.

Fabozzi, F. J. (1995). *Investment Management*. Jakarta: Salemba Empat.

Faza, M. F. and Erna, H. (2014). The influence of intellectual capital on profitability, productivity, and company value on banking companies listed in the Indonesia Stock Exchange (IDX). *Jurnal Ekonomi dan Bisnis Islam*, 8(2).

Fernandes, N. and Ferreira, M. A. (2007). The evolution of earnings management and firm valuation: A cross-country analysis. EFA 2007 Ljubljana Meetings Paper. Available at SSRN: https://ssrn.com/abstract=965636 or http://dx.doi.org/10.2139/ssrn.965636.
Gordon, M. J. (1963). Optimal investment and financing policy. *Journal of Finance*.
Gozali, A. and Saarce E. H. (2014). The influence of intellectual capital on financial performance and company value,especially in financial industry and mining industry listed in Indonesia Stock Exchange Year 2008–2012. *Business Accounting Review*, 2(2).
Guthrie, J. and Petty, R., Intellectual capital: Australian annual reporting practices,. *Journal of Intellectual Capital*, 1 (3), August, 2000, 241–51.
Halim, A., Basri, H., & F. (2016). The influence of intellectual capital on profitability and the impact on the stock prices of financial-sector companies listed in the Indonesia Stock Exchange (IDX). *Jurnal Bisnis dan Ekonomi*, 23(2).
Harijanto, V. A. and Mildawati, T. (2017). The influence of free cash flow on company value mediated by earning management. *Jurnal Ilmu dan Riset Akuntansi*, 6(1).
Indriantoro, Nur and Bambang Supomo. (2009), Business Research Methodology, BPFE, Yogyakarta.
Jensen, M. and Meckling, W. (1976). Theory of the firm: Managerial behavior, agency,and ownership structure. *Journal of Financial Economics*, 3, pp.305–360.
Kouki, M. and Guizani, M. (2009). Ownership structure and dividend policy evidence from the Tunisian Stock Market. *European Journal of Scientific Research*,25(1), pp. 42–53.
Lintner, J. (1956). Distribution of incomes of corporations among dividends, retained earnings and taxes. *American Economic Review*, 46, pp. 97–113.
Litzenberger, R.H. and Ramaswamy, K. (1979). The effect of personal taxes and dividends on capital asset prices: Theory and empirical evidence. *Journal of Financial Economics*, 7, pp. 163–195.
Masbachah. (2017). The influence of corporate social responsibility (CSR), independent commissioner, leverage, profitability and size toward tax avoidance. Undergraduate Thesis. Fakultas Ekonomi, Sultan Agung Islamic University.
Mayarina, N. A. and Mildawati, T. (2017). The influence of financial ratio and FCF toward company value with dividend policy as the moderator. *Jurnal Ilmu dan Riset Akuntansi*, 6(2).
Naini, D. I. and Wahidahwati. (2014). The influence of free cash flow and institutional ownership on debt policy and company value. *Jurnal Ilmu dan Riset Akuntansi*, 3(4).
Pambudi, N. M. and Andayani. (2017). The influence of intellectual capital and disclosure index on company value. *Jurnal Ilmu dan Riset Akuntansi*, 6(5).
Pertiwi, N. T. and Priyadi, M. P. (2016). The influence of profitability, managerial ownership, DER and FCF on company value through DPR. *Jurnal Ilmu dan Riset Akuntansi*, 5(2).
Pracihara, S. M. (2016). The influence of debt policy, managerial ownership, dividend policy, and company size toward company value. *Jurnal Ilmu Manajemen*.
Prastika, I. W. and Amanah, L. (2017). The influence of financial performance, dividend policy, investment decision and company size on company value. *Jurnal Ilmu dan Riset Akuntansi*, 6(7).
Putri, G. A. P. and Chabachib, M. (2013). Analysis of institutional ownership effect, free cash flow, and investment opportunity toward company value with debt policy as an intervening variable (case study of non-financial companies listed on IDX Year 2008–2011). *Diponegoro Jurnal of Management*, 2(2).
Riyadi, S. (2018). Analysis of manufacturing company's value. *Jurnal Sinar Manajemen*, 5(1).
Stewart, T.A. (1997) Intellectual capital: The new wealth of organizations. New York: Doubleday.
Yuskar., Novita, D. (2014). Analysis of intellectual capital's effects toward company value with financial performance as an intervening variable in Indonesia banking companies. *Jurnal Manajemen dan Bisnis Sriwijaya*, 12(4).

*Facing Global Digital Revolution – Nirmala Arum Janie,
Dwi Mulyaningsih & Wahyu Rachmawati (eds)*

Using diamond fraud analysis to detect fraudulent financial reporting of Indonesian pharmaceutical and chemical companies

Saifudin, Desy Rahmania Efendy & Dyah Nirmala Arum Janie
Universitas Semarang, Kota Semarang, Indonesia

ABSTRACT: This study aimed to detect fraudulent financial reporting using diamond fraud analysis. This rationalization was proxied by an audit and the capability to replace directors suspected of making fraudulent financial statements. The sample comprised 10 pharmaceutical and chemical sub-manufacturing companies listed on the Indonesia Stock Exchange (IDX) from 2012 to 2017. This research was conducted using multiple linear regression analysis. The results showed that external pressure variables measured by leverage ratios influence fraudulent financial reporting. Financial stability and change in terms of value, along with how a change of directors affects fraudulent financial statements, were also examined.

Keywords: fraud diamond analysis, fraudulent financial statements, financial stability, external pressure, financial targets

1 INTRODUCTION

Financial statements are a corporate communication tool regarding financial data or company operational activities. Financial statements reflect company performance within a certain period. Recognizing the importance of financial statements motivates managers to improve company performance. Unfortunately, not all company managers realize the importance of clean financial reports that are free of fraud (Yesiariani, 2016).

Huge fraudulent financial reporting cases continue to take place after the ENRON scandal in December 2001. One Indonesian incident was the case of PT. Kimia Farma, Tbk, in December 2001. The management reported a net profit of IDR 132 billion on December 31, 2001, audited by Hans Tuanakotta and Mustofa (HTM). However, the Ministry of State-Owned Companies (BUMN) and the Capital Market Supervisory Agency (Bapepam) considered that the net profit was too high and contained manipulation. After a re-audit on October 3, 2002, the 2001 financial statements of PT Kimia Farma were restated because auditors found a reasonably simple mistake. In the new financial statements, the profits made were only IDR 99.56 billion, IDR 32.6 billion (24.7%) lower than the initial profit reported. The error was due to IDR 2.7 billion in overstated sales in the raw material industry unit, IDR 23.9 billion in overstated inventory in the central logistics unit, and IDR 8.1 billion in overstated inventory and IDR 10.7 billion in overstated sales in the pharmaceutical wholesalers' unit. The researcher adopted the Annisya, Mafiana, Lindrianasri, and Asmarianti (2016) study, which aims to analyze the data, and found empirical evidence regarding the fraud risk factor according to diamond fraud analysis: pressure, opportunity, rationalization, and capability. This study used eight independent proxy variables: financial targets, financial stability, external pressure, nature of the industry, active monitoring, change in auditors, rationalization, and capability.

Based on this background, we can see that fraudulent financial statements cause a significant loss not only for companies, but also for the users of financial statements. The four conditions mentioned previously motivated the fraudulent financial reporting. Therefore, scholars conducted necessary research regarding fraudulent financial reporting in pharmaceutical and chemical subsector manufacturing companies in Indonesia. The pharmaceutical and chemical subsector was chosen due to the rapid development of this industry with a market that is sufficiently developed to allow fraudulent financial reporting.

2 LITERATURE REVIEW

The higher the total assets owned by a company, the more wealth it has. A change in the percentage of total assets indicates the occurrence of manipulation in financial statements. Therefore, the ratio of changes in total assets is a proxy for the financial stability variable (Annisya et al., 2016; Sihombing and Rahardjo, 2014), and showing financial stability has a significant positive effect on the risk of fraudulent financial reporting. Yesiariani (2016) shows that financial stability harms detection of fraudulent financial reporting.

H1: Allegedly asset change affects fraudulent financial reporting.

A change in the high percentage of debt indicates manipulation in financial statements. Therefore, we used the leverage ratio as a proxy for the external pressure variable. Yesiariani (2016) and Putriasih, Herawati, and Wahyuni (2016) show that external pressure has a significant positive effect on the risk of financial statement fraud. Annisya et al. (2016) reveal that external pressure has a significant negative effect on fraudulent financial reporting. Based on the description, this study proposes the following hypothesis:

H2: Allegedly leverage affects fraudulent financial reporting.

Pressure on the achievement of financial targets to get performance bonuses and to maintain existing company performance can raise the risk for fraudulent financial reporting. Chyntia and Puji (2016) and Annisya et al. (2016) prove that the return on assets (ROA) variable has a positive but not significant effect on fraudulent financial reporting. Sihombing and Rahardjo (2014) show that financial targets do not significantly influence financial statement fraud. Based on the description, this study proposes the following hypothesis:

H3: Return on assets affects fraudulent financial reporting.

A high ending inventory balance can also generate high profits. It is what triggers management to commit fraud on the inventory balance. So a percentage of change in total inventory can indicate a high level of fraud in financial reporting. Sihombing and Rahardjo (2014) state that the nature of the industry has a positive effect on fraudulent financial reporting. Ardiyani and Utamiingsih (2015) contend that the nature of the industry with a proxy stock (inventory) does not affect financial statement fraud. Based on the description, this study proposes the following hypothesis:

H4: Suspected inventory affects fraudulent financial reporting.

Auditors' opinions provide unqualified, descriptive language. To issue an opinion, an external auditor needs to identify and consider the risk factors that may cause the audit client to commit fraud. The existence of this opinion allows management to rationalize mistakes, because the auditor has tolerated the mistakes through the descriptive language in his opinion. Higher rationalization can therefore lead to fraud in financial reporting. Mafiana et al. (2016) also support Fimanaya and Syafruddin (2014) because the language in independent auditors' reports can be explanatory in some cases, such as reasonable opinions given in part based on other independent reports, additional information required by the Indonesian Accountants' Association, and certain other circumstances. Based on the description, this study proposes the following hypothesis:

H5: Explanatory, unqualified audit opinions affect fraudulent financial reporting.

A change of directors can indicate a company's effort to improve on the performance of previous directors by making changes to the composition of the board of directors or by recruiting new directors who are considered more competent. The more often companies make changes to directors, the higher their ability to commit fraud in financial statements. Chyntia and Puji (2016) conclude that changes in company directors have no effect in detecting fraudulent financial reporting. Annisya et al. (2016) and Sihombing and Rahardjo (2014) also show that changes in directors have no significant effect on the risk of fraudulent financial reporting. Based on the description, this research suggests the following hypothesis:

H6: Allegedly, a change of directors has an effect on fraudulent financial reporting.

3 METHOD

The population in this study comprised companies engaged in the manufacturing sector of the chemical and pharmaceutical subsectors listed on the Indonesia Stock Exchange (IDX) in 2012–2017. The researchers completed purposive sampling in order to get a representative sample according to the specified criteria.

4 RESULTS AND DISCUSSIONS

Assets change was not proven to affect fraudulent financial statements. The results of this study were consistent with Yesiariani (2016), but failed to prove Annisya et al. (2016) or Sihombing and Rahardjo (2014). The fundamental reason this research was insignificant is that the greater the change in company assets as a proxy for financial stability, the smaller the practice of fraudulent financial reporting. It happened because a large asset change caused the company to get public attention and the news spread quickly. According to Yesiariani (2016), companies have an excellent level of supervision carried out by a board of commissioners to monitor and control management policy. They are directly responsible for business functions such as finance so that even though management faces pressure when there is a threat to financial stability, economic, industrial, and operating conditions will not affect the occurrence of fraudulent financial reporting.

Leverage is proven to affect fraudulent financial reporting. The results of this study were consistent with Yesiariani (2016) and Sihombing and Rahardjo (2014). However, this study contradicted Annisya et al. (2016) and Susanti (2014). The target financial statement has no significant effect, so it cannot be used to detect fraudulent financial reporting. The results of this study were in line with those of Annisya et al. (2016), Sihombing and Rahardjo (2014), and Yesiariani (2016). However, they were contrary to Putriasih et al. (2016).

The fundamental reason this research was not significant is because the higher a company's ROA, the smaller the practice of fraudulent financial reporting. This is because every increase in the ratio of ROA does not become a pressure on the management of the company, because an increase in operational quality accompanies it. A decrease in company profitability can occur due to an unpredictable crisis that hit the industry (Sihombing and Rahardjo, 2014). So even though management faces pressure because it is unable to meet its financial targets, the occurrence of fraudulent financial reporting will remain unaffected.

The nature of the industry proxied by inventory has no significant effect, meaning it cannot be used to detect fraudulent financial reporting. The results of this study corroborate Annisya et al. (2016), Yesiariani (2016), and Susanti (2014). However, this research is not in line with Sihombing and Rahardjo (2016). The absence of a significant influence on the nature of the industry is because any increase or decrease in the inventory change ratio does not compel the company management to commit financial statements fraud (Annisya et al., 2016). Pharmaceutical companies usually produce large quantities of goods to meet high market demand, so changes in inventory cannot be used as an opportunity for management to manipulate financial statements.

Rationalization proxied by unqualified audit opinion with explanatory language cannot be used to detect fraudulent financial reporting. The reasons underlying these insignificant results are the addition of explanatory language in the independent auditor's report in part based on other independent reports, additional information required by the Indonesian Accountants' Association, and certain other conditions. This given opinion under certain circumstances requires the auditor to add explanatory paragraphs in the audit report, although it does not affect the unqualified opinion expressed by the auditor (Annisya et al., 2016).

A capability that is proxied by a change of directors does not have a significant effect, meaning that it cannot be used to detect fraudulent financial reporting. The results of this study are in line with Annisya et al. (2016), Yesiariani (2016), and Sihombing and Rahardjo (2014). However, it is contrary to Putriasih et al. (2016). The fundamental reason this research was not significant is because the supervision of a board of commissioners is related to the performance of each director. Also, a change in directors made it possible to motivate management to perform better than before because of the recruitment of more competent directors (Annisya et al., 2016).

5 CONCLUSION

Because of the limitations of this study, future studies are expected to add proxy variables from diamond fraud analysis so that the scope of the research variables becomes wider. Future studies might find proxies for financial stability, external pressure, financial targets, nature of the industry, rationalization, and capability because the proxies for these variables are changes in assets, leverage, returns on assets, inventory, audit opinion, and change of directors and are still challenging topics to examine. Subsequent research can increase the number of sample categories in order to predict cases of fraudulent financial statements in other categories, such as nonfinancial and financial companies. Further research should also expand the observation period so that it can better describe fraudulent financial statements. Future studies may find other variables to measure rationalization and capability, and use other methods such as interviews to obtain more accurate and diverse research results.

REFERENCES

Ardiyani, S. and Utaminingsih, N. S. (2015). Analisis determinan financial statement melalui pendekatan fraud triangle. *Accounting Analysis Journal*, 4 (1), pp. 1–10.

Fimanaya, F. and Syafruddin, M. (2014). Analisis faktor-faktor yang mempengaruhi kecurangan laporan keuangan. *Diponegoro Journal of Accounting*, 3 (3), pp. 1–11.

Putriasih, K., Herawati, N. N. T., and Wahyuni, M. A. (2016). Analisis fraud diamond dalam mendeteksi financial statement fraud: Studi empiris pada perusahaan manufaktur yang terdaftar di Bursa Efek Indonesia (BEI) tahun 2013–2015. *Jurnal Ilmiah Mahasiswa Akuntansi Undiksha*, 6 (3), pp. 1–22.

Sihombing, K. S. and Rahardjo, S. R. (2014). Analisis fraud diamond dalam mendeteksi financial statement fraud: Studi empiris pada perusahaan manufaktur yang terdaftar di Bursa Efek Indonesia (BEI) tahun 2010–2012. *Diponegoro Journal of Accounting*, 3 (2), pp. 1–12.

Susanti, Y. A. (2014). Pendeteksian kecurangan laporan keuangan dengan analisis fraud triangle. Thesis, Universitas Airlangga.

Ujiyantho, M. A. and Pramuka, B. A. (2007). Mekanisme corporate governance, manajemen laba dan kinerja keuangan. *Simposium Nasional Akuntansi X*, pp. 1–26.

Yesiariani, M. and Rahayu, I. (2016). Analisis fraud diamond dalam mendeteksi financial statement fraud (studi empiris pada perusahaan LQ-45 yang terdaftar di Bursa Efek Indonesia tahun 2010–2014. *Simposium Nasional Akuntansi XIX*, pp. 1–22.

Facing Global Digital Revolution – Nirmala Arum Janie, Dwi Mulyaningsih & Wahyu Rachmawati (eds)
© 2020 Taylor & Francis Group, London, ISBN 978-0-367-33912-8

A study of the effectiveness of e-commerce adoption among Small and Medium-sized Enterprises (SMEs) providing postnatal care services: A case study in Selangor, Malaysia

I. Masbuqin & S.M. Sharif
Universiti Teknikal Malaysia, Melaka, Malaysia

A. Isa
Universiti Teknologi MARA (UiTM), Machang, Malaysia

A.Y.M. Noor
Universiti Kebangsaan Malaysia (UKM), Bangi, Malaysia

A.Z. Samsudin
Universiti Teknologi MARA (UiTM), Machang, Malaysia

ABSTRACT: The research covered in this article aimed to provide an insight into factors affecting business-to-business e-commerce (EC) adoption and implementation in small to medium-sized enterprises (SMEs) providing postnatal care services. However, EC adoption by SMEs in developing countries has faced many challenges that we have not adequately addressed due to the complex nature of EC adoption in such countries. We based this research on an extensive literature review, and focused on proposing a theoretical model of technological, environmental, and organizational factors (TOE) influencing EC adoption and implementation. Thus, our first and second research goals were to identify the characteristics and perceptions of EC adoption among SME managers working in the postnatal care industry. Finally, this research also covered critical studies, as it tried to provide organizations with key strategies for creating awareness of EC platforms among SMEs involved in postnatal care services. The results of the empirical research provide an indication that SMEs in the postnatal care services industry are interested in adopting and conducting business-to-business EC. This article both contributes to our understanding of the factors affecting business-to-business EC adoption and implementation in SMEs and provides interesting perspectives from the postnatal care services industry.

Keywords: e-commerce, small to medium-sized enterprise (SME), e-commerce adoption, postnatal care services industry

1 INTRODUCTION

Malaysian women customarily stay at home after giving birth and observe a month-long confinement period, and other traditional postnatal practices are also widely used in Malaysia. The remarkable experience of Malaysia in lowering maternal mortality reflects a robust strategic plan to improve maternal and child well-being. When all is said and done, Malaysia has seen noteworthy enhancements in postnatal maternal and child well-being. Even before the Safe Motherhood Initiative, the claimed maternal mortality (MMR) ratio increased from about 280 to 141 per 100,000 live births between 1957 and 1970 (MGD5, 2004).

According to the 2013 estimates of maternal mortality by the World Health Organization (WHO, 2014), 99% (286,000) of worldwide maternal deaths occurred in developing nations.

Maternal death statistics were proposed to include slow deaths within one year postnatal, as the risk of death increases for up to six months after birth (Høj et al., 2003). Wickramansinghe (2005) states that an exchange of information is necessary and common because knowledge is allocated among many organizations and companies that recognize an expanding need to cooperate. Because of their role in promoting economic growth and innovation, small and medium-sized enterprises (SMEs) are an essential sector of the economy.

2 CONCEPTUAL BACKGROUND AND A RESEARCH MODEL

Delone and Mclean (2004) inspect information exchanges under the first D&M model and its augmentations, and have added administration quality to their refreshed information services (IS) achievement model. Tornatzsky and Fleischer (1990) build up the technological, organizational, and environmental (TOE) processes for technological innovation. In their book, the TOE system alludes to the structure for development appropriation and execution with regards to an organization, and it is applied to three unique contexts. The use of TOE in the SME context has been extended to the activity of the chief executive officer (CEO) (Thong, 1999), who a significant part of the time is the first CEO as well as the owner of a business. The TOE system may be useful in the SME setting in cases when it can fuse the whole scope of management (Robertson, 2010).

2.1 *Characteristics and perceptions of e-commerce (EC) adoption among SME managers working in the postnatal care industry*

Proposition 1: A significant relationship exists between managerial perception and characteristics and EC adoption.

The adoption of EC in an organization leads to changes that require the development of a strategic perspective and the ability to deal with these changes. When these organizations succeed, it is because they have focused on leadership that facilitates the appropriate changes and creates a conducive climate for EC. This study supports Cope and Waddell's (2004) argument that the specific characteristics required to carry out an EC transition are vision, inclusivity, risk-taking, accessibility, forward-thinking, openness to change, commitment, determination, and communication capability.

Proposition 2: A significant relationship exists between the TOE contexts and EC adoption.

2.2 *Technological context*

The technological framework includes all technologies that are relevant to a company, along with technologies already used in the company and others on the market but never used in the company. According to Petter and McLean (2009), IS literature indicates that EC/IS technological features and the use of and satisfaction with IS are essential factors for success, including information, the system, and the quality of service. The quality of the information we can characterize as the nature of the data given to an association by its IS concerning practicality, precision, dependability, significance, and data fulfilment (Byrd et al., 2006). Little has been done to comprehend the association between the nature of EC applications' data and the utilization of EC in a business-to-business (B2B) situation. For instance, extremely advanced knowledge acquisition methods are typical of customer feedback systems, data mining, company intelligence, and cooperation with partners and study organizations (Alias, Mansor, Rahman, Ahmad, and Samsudin, 2018).

2.3 *Organizational context*

We frequently distinguish support from top management in an organization as a significant component in any technological adoption inside an organization. Oliveira and Martins (2010) opine that management support and interest help to bring to light an organization's innovative focal points through formal communication, give the required innovation framework, and support technology that improves the probability of advantage acknowledgment and diminishes the organization's protection from change. We, therefore, follow the findings of previous successful research (Sabherwal et al., 2006) that shows that management support has a constructive and direct impact on user satisfaction and system utilization. In the meantime, successful IS adoption requires knowledge sharing, training, and a higher level of skills among staff members who are also IS users (Egbu, Hari, and Renukappa, 2005). Individual and organizational capabilities cannot be affected without disturbing anyone else's work process (Sharif, Ahamat, et al., 2018). This is especially essential because it requires not only understanding but also social skills (Nur Fauzi, Usodo, and Subanti, 2017). Therefore, *knowledge sharing* can be defined as communication between two or more people involved in providing and acquiring knowledge that can take many forms, using either verbal or nonverbal mechanisms, and with or without the use of technology (Isa et al., 2016).

2.4 *Environmental context*

Regulatory support also plays a vital role in influencing the adoption of technologies by organizations (Tan and Ouyang, 2006). Wahid (2007) states that regulatory support also plays an essential role in providing the necessary national information and communications technology (ICT) infrastructure, such as a reliable Internet connection, with reasonable speed and appropriate technical standards. Government backing can come through encouraging SME activities, providing financial and innovative help, improving the EC framework, and authorizing laws that are friendly to online business (Scupola, 2003). Government facilitates the exchange of business development in a legitimately organized way for public use (Sharif, Nizam, et al., 2018). Therefore, we used TOE as a general framework for the study of various influential EC adoption factors.

3 FINDINGS AND DISCUSSION

There seemed to be an absence of apparent connection between expected results of the adoption of EC projects and organizational objectives. Specified organizational goals seem to have a crucial role to play in organizing and conducting investment, and also in the priorities assigned to various information technology (IT) investment initiatives (Mirani and Lederer, 1993). The goals for organizations to adopt EC systems have varied. Rogers (2003) demonstrates a time frame for the adoption of a new technology/idea where few adopters arrive at the beginning and with time, the percentage of users increases. Upper-management commitment across the implementation phase was crucial to the success of EC IT investments (Power, 2004). All of the organization representatives interviewed stated that top executives provided adequate management leadership and perhaps even the necessary organizational commitment to implement IT investments in EC. The interaction of components and how they decided the extent of EC adoption in SMEs depended on the conditions of each organization.

Some organizations have profited enormously from EC adoption, for example, C1Alpha, some of which have been reasonably influenced by C3Star, and others that are at much lower dimensions of the C2Puma advantage. It is consequently essential to comprehend what helps SMEs to accomplish these positions. The discussion depends on these dimensions of utilization in SMEs. For C1Alpha, the extent of EC adoption has been affected by elements portrayed as high-level drivers, for example, the potential customers available in Selangor and the inventive and innovative abilities of managers and existing ICT applications – for example, email, the Internet, and a dynamic site giving guidelines for payment.

Proposition 3: A significant relationship exists between key strategies suggested to organizations for creating awareness and EC.

The communication of these interior variables with the environmental context gives C1Alpha an incredibly uplifting position. The social component of EC transactions has enormously affected the nature of these exchanges, as the gatherings confide in one another and products are purchased. What is more, when idospa.my utilized a third-party agent in the C3Star models, the company added to its capacity to advertise its products and services adequately in the worldwide market. C3Star's outcomes verify past studies by Ray and Ray (2006), which demonstrate that SMEs could, with minimal effort, exploit third-party web services in order to offer strategic incentives in money generation and higher organizational benefits.

4 CONCLUSION AND FUTURE RESEARCH

In this research, we left a related or nearly similar improvement on EC adoption in the postnatal care services industry for future study. Future research should consider the allocated time frame for completing the research. Another postnatal care sector in the Selangor region and also in Malaysia can provide more accurate and efficient data. There are, however, best practices from the research that organizations can use (Husin, Yunus, and Samsudin, 2017).

At long last, EC adoption in SMEs is expansive and covers different aspects, for example, promotion, organizational behavior, and others. Employee satisfaction is a significant factor in any corporation's achievement. The compensation of the staff must be sufficient to satisfy the requirements of a higher living standard (John and Vikitset, 2019).

ACKNOWLEDGMENT

The author would like to thank Universiti Teknikal Malaysia Melaka (UTeM) and the Centre for Technopreneurship Development (CTED) for their support in obtaining materials for this study and research.

REFERENCES

Delone, W. H. and Mclean, E. R. (2004). Measuring e-commerce success: Applying the DeLone & McLean information systems success model. *International Journal of Electronic Commerce*, 9 (1), pp. 31–47.

Husin, S., Yunus, A. M., and Samsudin, A. Z. H. (2017). Empowering technical education institutes through harvesting tacit knowledge: An empirical study at Kolej Kemahiran Tinggi MARA Kemaman, Malaysia. *International Journal of Academic Research in Business and Social Sciences*, 7 (11), pp. 826–848.

Oliveira, T. and Martins, M. F. (2010). Understanding e-business adoption across industries in European countries. *Industrial Management & Data Systems*, 110 (9), pp. 1337–1354.

Ray, A. W. and Ray, J. J. (2006). Strategic benefits to SMEs from third party web services: An action research analysis. *Journal of Strategic Information Systems*, 15 (4), pp. 273–291.

Scupola, A. (2003). The adoption of Internet commerce by SMEs in the south of Italy: An environmental, technological and organizational perspective. *Journal of Global Information Technology Management*, 6 (1), pp. 52–71.

Sharif, S., Ahamat, A., et al. (2018). University intellectual property commercialization: A critical review of literature. *Turkish Online Journal of Design Art and Communication*, 8, pp. 874–886.

Sharif, S., Nizam, N., et al. (2018). Role of values and competencies in university intellectual property commercialization: A critical review. *Turkish Online Journal of Design Art and Communication*, 8, pp. 887–904.

Wahid, F. (2007). Using the technology adoption model to analyze internet adoption and use among men and women in Indonesia. *Electronic Journal of Information Systems in Developing Countries*, 32 (1), pp. 1–8.

World Health Organization (2010). WHO technical consultation on postpartum and postnatal care. http://Whqlibdoc.Who.Int/Hq/2010/WHO_MPS_10.03_Eng.Pdf

World Trade Organization (2013). E-commerce in developing countries: Opportunities and challenges for small and medium-sized enterprises.

Facing Global Digital Revolution – Nirmala Arum Janie,
Dwi Mulyaningsih & Wahyu Rachmawati (eds)

Description of quality culture and corporate performance. Case study: Indonesia SMEs

Ayi. Tejaningrum
STIE Ekuitas, Bandung, Indonesia

ABSTRACT: The objective of this research was to describe the profile of quality culture and corporate performance by analyzing 264 creative industry small to medium-sized enterprises (SMEs) in nine cities in West Java, Indonesia. A questionnaire was used as a research instrument. The SMEs' representatives were asked to demonstrate the practices of quality culture and corporate performance. Quality culture variables were analyzed along seven dimensions: coordination between functions, customer orientation, competitor orientation, employee role, leadership role, transfer of knowledge and skills, and problem solving. For company performance, the examined indicators were business duration, employee absorption, and business change. The results of the study showed that the quality culture is very much determined by the commitment of the leader; SMEs with high company performance have a high-quality cultural dimension of leadership commitment and consumer orientation. The weakness of SMEs is their inability to transform knowledge and skills among members of organizations.

Keywords: corporate, performance, quality culture, SMEs

1 INTRODUCTION

Culture is closely related to values and beliefs that are reflected in daily actions. The quality-development methods that have been known – for example, total quality management (TQM), Six Sigma, QFD, QCC, etc. – often do not have a significant impact on the development of product quality, because they are influenced by the individual corporate culture. In applying TQM in each country, different indicators are provided, such as those presented by Kureshi, N. I. et al. (2009)identifying the success factors for implementing TQM in Pakistan, which differ from those carried out by Majumdar (2016), who implemented TQM in India, as well as research that Munizu (2013) carried out in Indonesia. History shows that TQM succeeded in Japan because it was influenced by Japanese culture, which was very thick with orientation to processes compared to results.

Academics and practitioners agree that quality is related to customer satisfaction. Consumers will take action if they get high satisfaction from product quality, including buying back (becoming a customer), buying other products produced by the company, conveying their satisfaction to acquaintances, and, finally, generally they (consumers) will not heed competitors' promotions as much. Thus the company must strive to achieve high performance by producing quality products that meet consumer expectations. Therefore, quality must be a culture in small to medium-sized enterprises (SMEs). This study aimed to analyze in more depth how SMEs demonstrate the implementation of the quality culture that occurs and how it relates to company performance.

2 LITERATURE REVIEW

This research was conducted on Indonesian SMEs spread across nine cities or districts in the West Java province of Indonesia with a total sample of 264 SMEs by analyzing quality cultural profiles and company performance profiles. For matters relating to the concept of SMEs, refer to the Law of the Republic of Indonesia Number 20 of 2008 concerning Micro, Small and Medium Enterprises (MSEs). The law addresses micro-businesses, small businesses, and medium-sized businesses, with the main differentiator being company turnover and assets.

To understand the issues related to the quality culture, you must first understand the concepts of culture and quality. Robbins, Stephen P & Judges, Timothy A. (2010).defines *culture* as a shared system adopted by members who distinguish their organization from other organizational organizations. This shared system of meaning, if observed more closely, is the main set of characteristics valued by the organization. Goetsch, D. L. and Davis, S. (2014) is the manifestation of its underlying values and traditional culture. *Organizational culture* is the set of shared values, beliefs, and norms that influence the way employees think, feel, and behave in the workplace (Schein, E. H.: 2010). According to Nelson, D. L., & Quick, J. C. (2011), organizational culture performs four functions: gives members a sense of identity, increases their commitment, reinforces organizational values, and serves as a control mechanism for shaping behavior.

Quality culture (Goetsch and Davis, 2014) is an organizational value system that results in an environment that is conducive to the establishment and continual improvement of quality. It consists of values, traditions, procedures, and expectations that promote quality. Hofstede, G. (2001), contends that *organizational culture* refers to the collective programming of the mind that distinguishes the members of one organization from the members of another. Research conducted by Pettigrew, Woodman, and Cameron (2001), Porter and James L Heskett (1992) tates that culture will be highly reflected in three variables: administration, management, and problem solving.

Robin (2011) also explains that organizational culture is related to how employees perceive the characteristics of an organization's culture. Organizations comprise three types of culture: the dominant culture (the values shared by a majority of organization members), a subculture (a smaller culture within an organization, usually determined by department signs and geographical separation), and core values (the primary or dominant values received throughout the organization). Kreitner (2010) mentions that organizational culture is a set of assumptions that are implicitly divided and held by one group that determines how the organization is perceived, how it is thought about, and how it reacts to diverse environments. Edgar Schein (2010) combines the understanding of ideational school and adaptation, arguing that organizational culture is shared basic assumptions that a group has learned as it has solved problems of external adaptation and internal integration, and that have worked well enough to be considered valid and to be taught to new organization members. Furthermore, Schein (2010) states that culture is a pattern of basic assumptions groups share when solving problems of external adjustment and internal integration.

Goetsch and Davis (2014) state that *quality* is a dynamic condition that relates to products, services, people, or processes that meet or exceed expectations. Harvey and Green (1993) outline the nature of quality culture, which was seen, at the time, as a function of the manufacturing industry: a culture of quality is one in which everybody in the organization, not just the quality controllers, is responsible for quality. A central feature of such organizations is that each worker or team of workers is both a customer of and supplier to other workers in the organization. They form a chain of internal customers and suppliers.

Thus quality culture is how organizations take action in producing products that are in accordance with the wishes and expectations of consumers. In general, the variables defined in the concept of quality culture are the following: coordination between functions, customer orientation, competitor orientation, employee roles, leadership roles, problem solving, and the transfer of knowledge and skills.

The concept of organizational performance adopts what is conveyed by Dermawan (2011). Kaplan, R. S. and Norton, D. P. (1996) mention four indicators to assess organizational performance: finance, marketing, internal business, and human resources growth. In some conditions, considering results and processes becomes one of the references for analyzing organizational performance. In this study, with the focus on SMEs, the analysis of organizational performance was calculated from the absorption of labor and the duration of business was a major consideration because this concerns growth and sustainability. Turnover and profit and cost are not taken into consideration, due to the low number of SMEs who conduct their financial administration in writing.

3 MATERIAL AND METHOD

The population of this study comprised SMEs spread across nine cities or districts in West Java – Sumedang, Tasikmalaya, Cianjur, Bandung, Sukabumi, Majalengka, Garut, Purwakarta, and Cimahi – with 264 SME samples. A total of 141 are engaged in the food processing or culinary industry, 91 produce handicrafts, 29 work in the clothing industry, and the remaining 3 are involved in printing.

4 RESULT AND DISCUSSION

4.1 *Heading value of quality culture*

The results indicated that Indonesian SMEs generally have an advantage in the high role of leaders in ensuring products are in accordance with what is consumers expect; this condition can be understood because leaders in SMEs are also often the owners, and the sustainability of a company depends on the owner. The main weaknesses of Indonesian SMEs come in transferring knowledge and skills from the leadership to members of the organization; this occurs because the quality of the process of transferring knowledge to members of the organization is low.

4.2 *Quality culture and corporate life cycle*

Generally, SMEs still look for mutual cultural values among their members. At the initial stage, the spirit of achieving the best product quality is very high given that this is the stage when the product is introduced to the market. This quality culture reaches its peak when the organization reaches the age of 6–10 years. When SMEs pass the initial stages of the company life cycle, the spirit of collaboration enhances the quality culture. Owners of SMEs are direct company leaders, different from large companies so that the culture of quality is largely determined by the values, attitudes, norms, values, and beliefs of the leadership. When the company enters the third (3.88) and fourth life cycles (3.82), the quality of cultural values decreases, largely due to fatigue from the leader and owner. Organizational culture is largely determined by individuals, not systems.

4.3 *Quality culture and leadership commitment*

There is a strong correlation between leadership commitment and other quality culture dimensions. If the leadership commitment is not strong, which is indicated by a small value, then the overall value of the quality of culture is seen to decrease, as well as when the leadership commitment increases, the average value of the quality of the culture rises. This condition shows that the single majority, known as a one-man show in SMEs, is very strong. The progress and decline of a company is very much determined by the role of its leader. Leaders in SMEs provide the dominant force in forming a quality culture. This condition is in line with the theory of cultural formation, where the owner will contribute the values, beliefs, attitudes, and norms that predominantly regulate the culture (Robbins, A. Judge, Timothy: 2011) This condition is one of the main weaknesses of SMEs, where dependence on leadership is very high.

4.4 *Quality culture and business ownership*

The results of several interviews revealed why people choose a profession as an entrepreneur or a manager of an SME: business opportunities (27%), no chance to work as an employee in the company (42%), invitations from relations, friends, and family (8%), because of their skills (7%), an unsupportive work climate, or organizational value incompatibility and self-passion (6%), or to work in a family business (10%). In general, SMEs are formed from a background because there is no other choice to sustain life; usually this condition is experienced by those with an undergraduate education.

4.5 *Quality culture and total revenue*

Based on studies in the field, as many as 10–25% of SMEs formed because of inheritance. In general, the weakness of SMEs in building a quality culture is the low level of transfer of knowledge, with an average value of 3.17, while the highest value is leadership commitment. The following is the average value of quality culture for each UKM with different companies serving.

5 CONCLUSIONS AND SUGGESTIONS

Based on this study, some conclusions can be drawn: a quality culture is formed by seven dimensions, namely (a) coordination between functions, (b) customer orientation, (c) competitor orientation, (d) employee role, (e) leadership commitment, (f) problem solving, (g) transfer of knowledge and skill. Field analysis shows that cultural values are still low related to SMEs' ability to transfer knowledge and skills. The power behind quality culture is a high commitment from the leadership of the company; this can be understood because SME leadership is at the same time the ownership, so the tendency is to keep the company producing products that consumers like. Further analysis shows that the quality culture of a business built from the ground up is higher than that of family-owned companies. In general, quality culture reaches its peak when a company reaches the second stage of its life cycle, usually at the age of five to six years, and continues to decline as the company ages. This condition is caused because SMEs are managed almost 100% by their owners, so owners' fatigue is the cause of the decline in the company's quality culture. Other data show that when leadership commitment is high, the other dimensions of quality culture increase.

REFERENCES

Dermawan, W. (2011). *Corporate and Organizational Performance Management*. Jakarta: Erlangga.

Goetsch, D. L. and Davis, S. (2014). *Quality Management for Organizational Excellence: Introduction to Total Quality*. Seventh Edition. Pearson Education Limited.

Harvey, L. and Green, D. (1993). Defining quality. *Assessment & Evaluation in Higher Education*, 18(1), pp. 9–34.

Hofstede, G. (2001). Culture's consequences: Comparing values, behaviors, institutions, and organizations across nations. London: Sage.

Kaplan, R. S. and Norton, D. P. (1996). Using the balanced scorecard as a strategic management system. *Harvard Business Review*.

Kreitner, K. (2010). *Organizational Behavior*. New York: McGraw-Hill.

Kureshi, N. I. et al. (2009). Quality management practices of SME in developing countries: A survey of manufacturing SMEs in Pakistan. *Journal of Quality and Technology Management*, 5(2), pp. 63–89.

Majumdar, J. P. (2016). Causes of reluctance of Indian manufacturing SMEs to implement total quality management. *International Journal of Applied Research*, 2(2), pp. 126–134.

Munizu, M. (2013). Total quality management (TQM) practices toward product quality performance: Case of the food and beverage industry in Makassar, Indonesia. *IOSR Journal of Business and Management (IOSR-JBM)*, 9(2), pp. 55–61.

Nelson, D. L., & Quick, J. C. (2011). Understanding Organizational Behavior. Belmont, CA: Cengage South-Western.
Pettigrew, A. M., Woodman, R. W., and Cameron, K. S. (2001). Studying organizational change and development: Challenges for future research. *Academy of Management Journal*, 44(4), pp. 697–713.
Porter, Michael and James L Heskett (1992), Corporate Culture and Performance, New York, Free Press Page 78.
Robbins, Stephen P. & A. Judge, Timothy (2011). *Organizational behavior*. Fourteenth Edition. Pearson education. New Jersey 07458. 77–89.
Schein, E. H. (2010). *Organizational Culture and Leadership*. New York: Wiley.

Facing Global Digital Revolution – Nirmala Arum Janie,
Dwi Mulyaningsih & Wahyu Rachmawati (eds)
 ISBN 978-0-367-33912-8

Optimization of the strengths and weaknesses of MSMEs within the ASEAN Economic Community (AEC)

Wyati Saddewisasi, Sri Yuni Widowati, Djoko Santoso & Indarto
Universitas Semarang, Semarang, Indonesia

ABSTRACT: This study specifically aimed to describe: (1) strategic factors in micro, small, and medium enterprises (MSMEs) in dealing with the Association of Southeast Asian Nations (ASEAN) Economic Community (AEC); (2) the potency of business performance improvement in dealing with the AEC; and (3) a strategy of business performance development for MSMEs in dealing with the AEC. This research comprised an explorative descriptive study that analyzed the performance strategy of MSMEs. A survey was performed on five clusters of superior MSMEs in Semarang, namely *batik, food processing, handicrafts, tourism, and bandeng*. The population of this study consisted of MSME entrepreneurs who joined in the clusters in Semarang. Semarang was selected as the study location based on aspects of the city's economic development plan, especially in trade. Semarang's existing potential can be improved so that the city can compete in the AEC free market. The sampling of MSME entrepreneurs was decided purposively based on their homogeneity and on the real condition of MSME clusters in Semarang. The sample included 100 entrepreneurs with 20 people representing each of the five clusters. The processed data collected from the field were 78% from the sample, and were analyzed through a strengths, weaknesses, opportunities, and threats (SWOT) analysis. The results of this study showed that MSMEs can utilize their strengths to seize opportunities such as the establishment of clusters in order to build and manage relationships with the government and other parties. To minimize MSMEs' weaknesses, especially a low level of knowledge, a strategy can be created to exploit opportunities such as taking part in training and business development activities.

Keywords: strategy, strength, weakness

1 INTRODUCTION

The strength of micro, small, and medium enterprises (MSMEs) has been proven, especially in the period of economic crisis of 1997. The MSMEs were able to keep running their businesses and to reduce unemployment. According to Maria Jose et al. (2015), Fritsch and Storey (2014), and Mazzarol et al. (1999), between 90% and 99% of companies globally were small and medium enterprises with the majority consisting of small and even micro enterprises. The said small and medium enterprises played an important role in stability, employment, and economic development. Therefore, the MSMEs merit attention from all parties when analyzing the existing potential in each region, especially in this era of the Association of Southeast Asian Nations (ASEAN) Economic Community (AEC).

The era of the AEC, which was implemented in 2015, is both a challenge and an opportunity for MSMEs. The AEC provides integration in the form of a "free trade area," the removal of trade tariffs between ASEAN countries, and a free labor and capital market (Arisandy 2014). The AEC, therefore, greatly influences the economic growth and development of each member country.

To deal with the era of the free market, MSMEs of Central Java, in particular in Semarang, should build a strategy to compete with MSMEs of other ASEAN countries. This should be performed in order to solve all of the problems that occur in MSMEs. A basic problem faced by MSMEs of Semarang is poor business performance and an inability to compete in the free market. Semarang has five clusters of superior MSMEs: *batik, food processing, handicrafts, tourism, and bandeng.* Each of the clusters consists of approximately 50 members. Based on the aforementioned problems, it is necessary to conduct a study on strategy to utilize strengths and minimize weaknesses so that the MSMEs' business performance can be improved and they will be able to compete in the free market of ASEAN.

So far, the MSME problems in Semarang have been managed by the relevant agencies in both government and nongovernment circles. However, it is not yet known exactly which strategy is the most suitable to improve the business performance of MSMEs in Semarang. Therefore, the research questions were as follows: (1) What are the strategic factors in MSMEs? (2) To what extent do existing MSMEs have the potential to develop their business performance? (3) What strategies can MSMEs take to develop their business performance? Therefore, the formulation of the problem was: what is the suitable strategy to utilize strengths and minimize weaknesses for MSMEs in Semarang so they can better compete in the free market of the AEC?

In particular, the objective of this study was to describe: (1) strategic factors MSMEs can use in dealing with the AEC; (2) the potency of MSMEs' business performance improvement in dealing with the AEC; (3) a business performance development strategy for MSMEs in dealing with the AEC.

The study results can be applied to entrepreneurs in general, especially entrepreneurs included in the MSME category who are facing the AEC free market. Besides that, the government can use these study results in carrying out guidance concerning operating within the AEC economy. For academia, these study results can be used as a reference for the development of similar research in the future.

2 RESEARCH METHOD

This study used a descriptive exploratory approach in analyzing MSMEs' strategies in facing the AEC. This study comprised a survey of five clusters of superior MSMEs in Semarang, namely *batik, food processing, handicrafts, tourism, and bandeng.* The population of this study consisted of MSME entrepreneurs who joined the these five clusters in Semarang. The choice of Semarang as a research location was determined based on consideration of aspects of the city's economic development plan, especially in trade. The existing potential of the MSMEs in Semarang can be improved so that they can compete in the AEC free market. The sampling of MSME entrepreneurs was carried out purposively based on their homogeneity and on the real conditions of MSME clusters in Semarang. The total sample of entrepreneurs was 100 people with a processed questionnaire return rate of 78%. The entrepreneurs were chosen because they could find the right strategy to compete in the era of the AEC free market.

The steps taken in this study were as follows: (1) identifying strategic factors in facing the AEC; (2) analyzing these factors by the total score of internal and external factors using the internal–external matrix model. The attributes used in the questionnaire were obtained from the perceptions of MSME entrepreneurs, which generally included attributes of the internal and external conditions of the MSMEs.

The initial step of this study was a focus group discussion (FGD) with MSME entrepreneurs to look for strengths, weaknesses, opportunities, and threats (SWOT) regarding strategic factors in the MSME cluster. Primary data were collected through interviews with the entrepreneurs.

Data obtained from the study field were analyzed descriptively using the internal–external matrix model and SWOT analysis. Descriptive data analysis described the MSMEs' strategy in facing the AEC free market.

3 RESULTS AND DISCUSSIONS

Based on calculations using the table of Internal Strategic Factors Analysis Summary (IFAS), which consisted of strengths and weaknesses, from the strength side it can be seen that MSMEs have advantages compared to their competitors. In this case, clusters can establish relationships with the government and other parties. This is the strength of the clusters that have formed in Semarang and that are members of the Forum for Development and Employment Promotion (FEDEP), an organization facilitated by the government of Semarang. The clusters have used this power to take advantage of existing opportunities. Besides that, to minimize weaknesses, especially a low level of knowledge, a strategy can be created to take advantage of existing opportunities. In this case, this can be performed by participating in business development training.

After the analysis of internal strategic factors, researchers then analyzed external strategic factors, which consisted of opportunities and threats. The results are shown in the External Strategic Factors Analysis Summary (EFAS) table. Free trade and online marketing presented a great opportunity for MSMEs in the AEC era. On the other hand, the biggest threat to MSMEs was better-quality products from competitors, unstable raw material prices, and expensive labor costs. To seize these opportunities, MSMEs can establish cooperation with various parties to develop their businesses and expand their market share. On the other hand, to avoid these threats, MSMEs can organize or participate in business development training to improve business and product quality and to reduce costs. In addition, the ability to communicate in foreign languages, especially English, needs to be improved.

By using the internal–external matrix model, a company's position and strategies used in facing the AEC era can be determined. According to Boris and Fish (2014), Rangkuti's (2008) determination of company position was based on an analysis of the total score of internal and external factors, using the internal–external matrix.

Field findings showed the internal–external matrix, with a total IFAS score of 2.64 and a total EFAS score of 3.32. Therefore, growth is the appropriate strategy for MSMEs in Semarang. The MSMEs can pursue growth within the AEC in terms of sales, assets, profits, or a combination of these. To increase sales, MSMEs must be able to develop new products, improve product quality, or increase access to a broader market. Besides that, in determining prices, they must be able to compete with foreign and imported products. The MSMEs can also minimize costs so that they can increase profits. This is the most important strategy because MSMEs are in a position of growth under competitive conditions. Therefore, market share is expected to increase. If a company has not gained profit from large-scale production, then it will experience defeat in competition. Therefore, MSMEs can focus on certain profitable markets.

In addition to the growth strategy, MSMEs can carry out concentration strategies through horizontal integration by expanding markets, and through increasing production and improving technology through internal and external development by strengthening their own and external resources through collaboration with fellow cluster members and other similar MSMEs.

4 CONCLUSION

Based on the results and discussion of this study, we can conclude as follows:

The existing strategic factors in MSMEs include internal factors, which consist of strengths and weaknesses, as well as external factors, which include opportunities and threats to MSMEs in the AEC era. The strategy that MSMEs can carry out to deal with the AEC is growth through horizontal integration. Activities that can be performed in accordance with these objectives are: increasing sales, strengthening resources, organizing or participating in business development training, and improving skills in foreign languages, especially English.

In order to increase sales, MSMEs must be able to develop new products, improve product quality, and widen their access to a broader market. Besides that, in determining prices, they must be able to compete with foreign and imported products, and they must minimize costs in order to increase profits. The MSMEs must also strengthen their own and external resources through collaboration with fellow cluster and other similar MSMEs. The MSMEs can organize or participate in business development training to improve business and product quality as well as to reduce costs in order to be more efficient. Furthermore, foreign language ability, especially in English, must be improved.

REFERENCES

Arisandy, Yuni. (2014). *Kesiapan Koperasi-UKM Indonesia menatap era MEA*. Webside: https://www.antaranews.com/berita/436319/kesiapan-koperasi-ukm-indonesia-menatap-era-mea-2015

Boris, E. and Fish, J. N. (2014). "Slaves no more": Making global labor standards for domestic workers. *Feminist Studies*, 40(2), pp. 411–443, 487–489.

Fritsch, M. and Storey, D. J. (2014). *Entrepreneurship in a Regional Context: Historical Roots, Recent Developments and Future Challenges*. Abingdon: Routledge.

Maria Jose, R-G, et al. (2015). *Entrepreneurial Orientation and Performance of SMEs in The Services Industry*, Vol. 28, No. 2, pp. 194–212.

Mazzarol, T. *et al.* (1999). *Factors influencing small business start-ups: A comparison with previous research. International Journal of Entrepreneurial Behavior & Research*, 5(2), pp. 48–63.

Rangkuti, Freddy. (2008). "*Analisis SWOT Teknik Membedah Kasus Bisnis*", Jakarta: Penerbit PT Gramedia Pustaka Utama.

Facing Global Digital Revolution – Nirmala Arum Janie, Dwi Mulyaningsih & Wahyu Rachmawati (eds)

Top three causes of failure in corporate management: Employees' insights

Anna C. Bocar & Juliet Sophia Gliten
Gulf College, Muscat, Sultanate of Oman

Hendrati Dwi Mulyaningsih
Universitas Islam Bandung, Bandung, Indonesia
Research Synergy Foundation, Bandung, Indonesia

Ani Wahyu Rachmawati
International Women University, Bandung, Indonesia
Research Synergy Foundation, Bandung, Indonesia

ABSTRACT: As the title of this article suggests, this study aimed to research the top three causes of failure of corporate management based on the insights of employees. This study employed a descriptive research design, and it took place in one of the colleges in the sultanate of Oman. The evening class students of the subject college comprised the participants of this study. This study consisted of 12 indicators borrowed from an online article. Questionnaires were administered to 103 male and 52 female students, for which permission was obtained. The frequency and percentile distributions were employed to arrive at quantitative answers for the profile of the respondents. A weighted mean was used to identify the top three causes of failure in corporate management. The findings of this research revealed that 2 out of the 12 indicators as utilized in this study are never or not at all the reasons for corporate management's failures. In addition, another two of the indicators occasionally or sometimes cause a failure of corporate management. The other eight indicators are frequently or most of the time the reasons why corporate management failed. The top three reasons for corporate management failure are fear of competition from other employees, the "I" syndrome, and inability to organize details. In closing, managers need to be cautious specifically in bringing together a company's details, they must avoid claiming what is due to the employees, and they must prioritize the employees since employees are important factors for a company's success. Moreover, managers must act like mature executives and avoid insecurities when interacting with other employees.

Keywords: employees, insights, managers, failure, causes

1 INTRODUCTION

Managers occupy a big role in the efficacious performance of firms or companies. Good services and quality products are fruits of the exemplary efforts of managers and their employees. They help in the creation of a company's name and goodwill. However, in one way or another, employees are vigilant in monitoring the performance of the heads of their organization. In silence, they evaluate the actions of their managers. Every step managers take matters to the organization as well as to the employees. The success or failure of management impacts a company, its managers, and its employees. This study was conducted in order to document the top three causes of failure of corporate management from the viewpoint of employees.

2 LITERATURE REVIEW

Strategic corporate plans in terms of expansion, cost of production, financial strategies, strengthening productivity, managing financial expenditures, and cautious risk management must be present to avoid failures. Mbat and Eyo (2013) argue that the most evident source of corporate failure lies in the inadequacy and incompetency of managers. Turner (2005) states that "companies succeed because by chance or circumstances their internal capabilities ... match the opportunities in the environment at that particular time." Furthermore, he contends that the engagement of an organization's key stakeholders is necessary for its success. This would imply that the competency of managers is part of the organization's success. Managers need to show their commitments and capabilities to face internal and external challenges. Moore (2018) suggests that lack of competency in management can cause corporate failures. One or more mistakes in managing an organization will further lead to failure. Common mistakes in managing an organization or a corporation include lack of leadership abilities, failure to create essential guiding principles, becoming arrogant or complacent, devaluing the power of vision or lack of mutual vision, doubling down when problems threaten the new vision, and inability to create short-term successes. In addition, managers sometimes place importance on designations and titles rather than understanding and knowledge. According to Mbat and Eyo (2013), it is also important that employees be empowered to lead in all areas of their job and be trained to improve job performance.

2.1 *Statement of the problem*

The purpose of this study was to look into the causes of failure in corporate management. Specifically, this study sought to determine the top three causes of corporate management failure based on the insights of employees.

3 RESEARCH METHOD

This study employed a descriptive research design. This method involved a survey conducted in one of the colleges in the sultanate of Oman. Evening class students participated in this research. They were chosen to answer the questionnaires since they are employed and believed to be extremely familiar with and experienced in managerial activities. The indicators utilized in this study were borrowed from an online article written by Denny (2016). These indicators were used to determine the top three causes of failure in corporate management as perceived by the students as employees. The questionnaires were administered to 103 male and 52 female students. Before the administration of the questionnaires, an email seeking permission was sent to the head of the faculty of the subject department. After permission was obtained, the researchers collected the data through the administration of the questionnaires to the evening class students in the college. In order to better interpret the data, frequency and percentile distributions were employed to arrive at quantitative answers for the profile of the respondents. A weighted mean was used to identify the top three causes of failure in corporate management from among the 12 indicators utilized in this study.

4 RESULT AND DISCUSSION

4.1 *Gender profile of the respondents*

Based on the gathered data, the number of male respondents was greater (66.45%) than the number of female respondents (33.55%). In the sultanate of Oman, it is quite visible that there are more male students than female students. However, this observation is most common in the evening classes since the morning classes host more female students as compared to their male counterparts.

4.2 *Causes of failure*

Along with changes in an organization, employees' responsibilities change as well, and the unlimited skills and abilities of employees are needed to achieve desired organizational changes. Kotter (2010) recognizes that a company should develop a vision and that the process must work well with teamwork; it needs the application of creative thinking since in this way a vision can be created and result in the desired future.

In this study, as revealed in the data gathered from the insights of the respondents, it was shown that two of the indicators are never or not at all the reasons why corporate management has failed: an emphasize on titles instead of knowledge and expertise ($\mu = 1.70$), and an expectation of pay for what employees know instead of what they do ($\mu = 1.10$).

Moreover, the respondents disclosed that two of the indicators are sometimes or occasionally the cause of corporate management failure: lack of understanding of the destructive effects of a negative environment ($\mu = 2.10$), and disloyalty to colleagues that results in loss of respect ($\mu = 2.24$).

According to Mbat and Eyo (2013), corporations fail due to several reasons. They stress that the incompetency of corporate managers in the utilization of resources in the most effective manner will result in the reduction of the return on investments and an inability to pay dividends to shareholders; however, a corporation can protect itself from failure through good relations between the management and employees. Having a pleasurable work environment will result in productivity, and when everyone in the corporation is productive, profitability increases.

Furthermore, the respondents stated that eight of the indicators are frequently or most of the time the reasons why corporate management failed. Out of these eight, the last three were declared by the respondents as the top reasons why corporate management failed. The first five indicators are overindulgence (leniency), which destroys endurance and vitality ($\mu = 2.51$); lack of common sense ($\mu = 2.52$); lack of creative thinking in setting goals and creating plans ($\mu = 2.59$); emphasis on the "authority of management" ($\mu = 2.61$); and unwillingness to do what they would ask another to do ($\mu = 2.63$). The last three reasons are the fear of competition from other employees ($\mu = 3.00$); the "I" syndrome – claiming all the honors for the team's achievements ($\mu = 3.01$); and the inability to organize details ($\mu = 3.20$).

In an organization, competition with the chief executive officer (CEO) leads to fear and pressure (Moore, 2018). Similarly, in a situation where a manager lacks executive managerial maturity, corporate management will fail because no matter how high the degree of necessity in the company for an employee who possess high qualifications, an immature manager might not hire an applicant whom he views as his potential competitor (Kay, 2019). The manager's refusal to hire someone smarter than him will hinder the growth of the company since this employee might have expertise that in-house employees do not possess. The knowledge and skills of one employee are a great contribution for the enhancement of others. Companies always consider the attainment of their goals, and achieving the same needs teamwork. This signifies that the presence of combined actions of a group of employees in a company is important. The manager is judged based on the output of his employees; however, when the manager claims the credit for his employees' achievements, this will harm his reputation and image since the employees must be his priority and not himself (Borysenko, 2019). It is unthinkable that a manager lacks the ability to organize details since this is one of the essential functions of a manager in a company; thus, the respondents in this study positively expressed that it is the top reason why corporate management failed.

5 FINDINGS

Based on the data gathered by the researchers, it can be observed that 2 out of the 12 indicators as utilized in this study are never or not at all the reasons for corporate management's failures. In addition, another two of the indicators occasionally or sometimes cause the failure of corporate management. Meanwhile, the other eight indicators frequently or most of the time are the reasons why corporate management failed. The top three reasons for corporate management failure are fear of competition from other employees, the "I" syndrome, and inability to organize details.

6 CONCLUSION

As perceived by the respondents, it can be concluded that managers must keep their eyes on the organization of their companies' details. Managers must avoid claiming what is due to the employees, and instead must give their employees priority since they are one of the important factors that help a company to succeed. Moreover, managers must act like mature executives and avoid insecurities when comparing themselves against other employees.

REFERENCES

Borysenko, K. (2019). Ten things new managers need to know. www.forbes.com

Denny, R. (2016). 12 Major causes of failure in leadership. www.1000advices.com

Kay, A. (2019). Tribe Total Media. https://archive.triblive.com

Kotter, J. P. (2010). *Leading Change*. Kindle edition. Boston, MA: Harvard Business School Press.

Mbat, D. O. and Eyo, E. I. (2013). Corporate failure: Causes and remedies. *Business and Management Research*, 2(4), pp. 19–24.

Moore, E. M. (2018). An exploration of the causes of success and failure of managed change. PhD dissertation. Walden University.

Turner, I. (2005). Corporate failure. *Henley Manager Update*, 17(2), pp. 33–42.

Facing Global Digital Revolution – Nirmala Arum Janie,
Dwi Mulyaningsih & Wahyu Rachmawati (eds)
 ISBN 978-0-367-33912-8

Internal control disclosure of companies with the most active stocks on the Indonesia Stock Exchange in 2016

Weli, S.M. Kusumawati & J. Sjarief
Atma Jaya Catholic University of Indonesia, Jakarta, Indonesia

ABSTRACT: Information about internal control is valuable for stakeholders, and previous studies indicate that Indonesian corporations need to improve their disclosure of such information. The current regulations require only minimum internal control disclosure in company reporting: (1) financial and operational control, (2) legal and regulatory compliance, and (3) control system effectiveness. Our research aimed to illustrate the internal control disclosure practices of companies with the most active stocks by trading volume on the Indonesia Stock Exchange in 2016. Thirty-seven companies were selected randomly for our observation. Through content analysis, we found an average score of 60.14 for internal control disclosure in annual reports. The findings of this study contribute to literature on internal control disclosure in annual corporate reporting.

Keywords: Internal control, disclosure, annual report

1 INTRODUCTION

An internal control system (ICS) produces policies that are necessary for achieving corporate objectives. Effectiveness of an ICS depends on the participation of company personnel, particularly those involved in accounting procedures and policies. Business processes are carried out based on established principles. Therefore, proper implementation of ICS will provide reliability of financial reporting and reduce the risks of fraud and material misstatement.

Despite the need of shareholders, the entirety of ICS information is available and accessible to internal parties. External stakeholders require disclosure about how an organization manages its business. Reporting on internal control is crucial for public companies (Deumes and Knechel, 2008) because it provides evidence of good corporate governance for both shareholders and financial regulators. Internal control disclosure may improve the overall quality and reliability of corporate financial reporting. Corporations are obliged by law to periodically evaluate the effectiveness of their internal control over financial reporting and to reveal the findings of the evaluation in a management report (Bronson, Carcello, and Raghunandan, 2006).

In Indonesia, regulations govern disclosure of internal control, albeit not as detailed as the SOX Act of 2002. This matter falls under the jurisdiction of the Financial Services Authority (FSA) – known in Indonesia as the Otoritas Jasa Keuangan. The FSA Circular Letter No. 30/SEOJK.04.2016 regulates internal control disclosure, specifically in the section about corporate governance. The FSA establishes that corporations must disclose internal control information about: (1) financial and operational control, (2) legal and regulatory compliance, and (3) review of the effectiveness of the control system. However, the regulations proved inadequate because they do not provide detailed information regarding content and format of the disclosure. This ambiguity leads to diverse or nonuniform disclosures by public companies. For the time being, stakeholders in Indonesia are provided assurance only through financial statements, a statement of management's responsibility, a management report on internal control over financial reporting, and letters on attestation engagements. Based on our given introduction, we formulated our research question as follows: what is the contemporary practice of corporate internal control disclosure in Indonesia?

2 LITERATURE REVIEW

For years, debate has taken place over the costs and benefits of disclosing internal control information in corporate reporting (Owusu-Ansah and Ganguli, 2010). Generally, previous research has discussed: (1) the reporting of internal control weaknesses or adequacy (Ashbaugh-Skaife, Collins, and Kinney, 2007; Doyle, Ge, and McVay, 2007; Ge and McVay, 2005), and (2) voluntary disclosure of internal control (Bronson et al., 2006; Deumes and Knechel, 2008; Michelon, Beretta, and Bozzolan, 2009). In Indonesia, internal control disclosure remains voluntary, which makes the extent of disclosure a topic worthy of investigation.

Hao and Rainsbury (2016) examine internal control disclosure of Chinese-listed companies subsequent to the central government's ruling of mandatory disclosure of management's assessment of the effectiveness of corporate internal controls and auditors' opinion. Seventy-three percent of the companies investigated were found to possess weak internal control and low quality of information disclosure. In Indonesia, several studies on internal control have been conducted and have found that the overall extent of internal control disclosure is low (Sjarief and Weli, 2016; Weli and Sjarief, 2017, 2018). This is potentially caused by deficiencies in current regulations, which do not dictate what information companies must disclose in corporate reporting. The voluntary nature of disclosure brings about varying reporting by companies.

This research formed a part of a series of studies on disclosure of internal control systems by publicly traded Indonesian corporations. Our current investigation differed from preceding ones (Sjarief and Weli, 2016; Weli and Sjarief, 2017, 2018) regarding the scoring or assessment of internal control information. This study observed public companies listed on the Indonesia Stock Exchange with the most active stocks by trading volume in 2016. Our primary aim was to obtain an overview of internal control disclosure practices by companies with similar characteristics.

3 RESEARCH METHOD

The population in this study comprised companies with the most active stocks by trading volume on the Indonesia Stock Exchange in 2016. The stocks were selected through simple random sampling and the collected data were analyzed using descriptive statistics. We measured internal control disclosure by adapting previous instruments (Leng and Ding, 2011; Van de Poel and Vanstraelen, 2011) to applicable regulations, specifically the Financial Services Authority Circular Letter No. 30/SEOJK.04.2016 concerning the format and content of annual reports of listed or public companies. We evaluated the effectiveness of internal control of our selected companies by adhering to the five components of the 2013 COSO Framework: Control Environment, Risk Assessment, Control Activities, Information and Communication, and Monitoring. Thirty-two items were used to calculate the internal control disclosure index.

4 RESULT AND DISCUSSION

After excluding financial firms, we had a population of 45 public companies with the most active stocks by total trading volume. Five companies were then removed due to incomplete data. We were left with 40 companies. We determined our sample size using Slovin's formula and ended up with 37 observations, selected randomly. Table 1 presents the mean score for internal control disclosure of the 37 sampled companies. We found an overall score of 60.14, indicating that not every company has disclosed internal control information in its respective annual reports. Out of the five COSO framework components, Monitoring had the highest score at 65.25 and Control Activities had the lowest at 56.37. Eighty percent of the companies have disclosed information about internal control evaluation by the audit committee, corporate governance structure, job description and duties of internal auditors, and

Table 1. Internal control disclosure score.

Description	Mean
Discussion about internal controls in the board of commissioners' report	45.95
Structure of corporate governance	83.78
Code of ethics	54.05
Corporate culture	51.35
Dissemination of code of conduct	56.76
Code of conduct enforcement efforts	51.35
Control Environment component	57.21
General description of risk management: Risk management goals	37.84
General description of risk management: Type of risks	37.84
General description of risk management: Risk management	89.19
Potential fraud taken into account by the organization	56.76
Acknowledgment of changes that significantly affect internal control	48.65
Acknowledgment of internal and external business risks	100.00
Risk Assessment component	61.71
Description of internal control over financial reporting	24.32
Description of internal control over financial reporting: Purpose of internal control	56.76
Description of internal control over financial reporting: Internal control policy	75.68
Description of internal control over financial reporting: Roles and responsibilities	43.24
Professional certification of the company's internal auditor	48.65
Accountability structure: Internal auditor's report to the audit committee, rather than to the CEO	62.16
Brief description of the internal audit department's job description and tasks during the financial year	83.78
Control Activities component	56.37
Presence of a medium for communicating information to internal stakeholders	24.32
Existence of a company website for communicating information to the public	100.00
Implementation of whistleblowing system	67.57
Description of the policy and procedures for managing complaints	67.57
Protection for complainants	59.46
The outcome of the handling of complaints	40.54
Information and Communication component	59.91
Management's evaluation of internal control's effectiveness	54.05
Commissioners' evaluation of internal control's effectiveness	70.27
Audit committee's evaluation of internal control's effectiveness	81.08
Examination and discussion of the effectiveness of internal control by the audit committee and internal auditors	72.97
Continuous improvement of the internal control system	45.95
Commissioners' assessment of the effectiveness of the risk management system	64.86
Management's assessment of the effectiveness of the risk management system	67.57
Monitoring component	65.25
Average score for internal control disclosure	60.14

risk management methods. However, 75% of the companies have yet to disclose their internal control systems, or even to own an internal communication medium. Sixty-two percent of the companies also have not revealed their risk management objectives and the types of risks they face. Fifty percent of the companies have not divulged the following information: discussion of internal control in commissioners' reports, changes in internal control, detailed description of internal control, internal control responsibilities, professional certification of internal auditors, whistleblowing outcomes, and continuous improvement of the internal control system.

In general, we found that internal control information has not been fully disclosed by public companies. Our findings corresponded to those of previous empirical studies.

5 CONCLUSION

Internal control disclosure in the annual reports of Indonesian public corporations is still limited. Sixty percent of the companies receive an average disclosure score of 60. This indicates that many items have not been widely disclosed, e.g., description of internal control over financial reporting, internal control responsibilities, existence of media for communicating information to internal stakeholders, outcome of complaint management, and an overview of corporate risk management. Therefore, we can conclude that Indonesian companies act on the FSA Circular Letter No. 30/SEOJK.04.2016 in a diverse manner; they comply only with specific regulations. The disclosed information includes (1) the use of web services, and (2) the identification and acknowledgment of changes that significantly affect the internal control system. Disclosures of other information are yet to be regulated by the FSA. Our research findings contribute to the literature on internal control disclosure. However, we were also subjected to limitations. The instrument for measuring ICS disclosure was adapted from previous studies, which were conducted in countries that mandate internal control disclosure. Subsequently, we adjusted the instrument with respect to applicable FSA regulations, thereby disregarding the standpoint of the public or listed companies. Future researchers could address this by considering internal company policies, e.g. what information should be announced to the public, and vice versa. This allows for a more symmetrical exchange of information between the company and the public. In conclusion, further investigations should focus on the needs and perceptions of financial statement users about the information disclosed in annual reports.

REFERENCES

Ashbaugh-Skaife, H., Collins, D. W., and Kinney Jr., W. R. (2007). The discovery and reporting of internal control deficiencies prior to SOX-mandated audits. *Journal of Accounting and Economics*, 44(1–2), pp. 166–192.

Bronson, S. N., Carcello, J. V., and Raghunandan, K. (2006). Firm characteristics and voluntary management reports on internal control. *Auditing: A Journal of Practice & Theory*, 25(2), pp. 25–39.

Deumes, R. and Knechel, W. R. (2008). Economic incentives for voluntary reporting on internal risk management and control systems. *Auditing: A Journal of Practice & Theory*, 27(1), pp. 35–66.

Doyle, J., Ge, W., and McVay, S. (2007). Determinants of weaknesses in internal control over financial reporting. *Journal of Accounting and Economics*, 44(1–2), pp. 193–223.

Ge, W. and McVay, S. (2005). The disclosure of material weaknesses in internal control after the Sarbanes-Oxley Act. *Accounting Horizons*, 19(3), pp. 137–158.

Hao, G. and Rainsbury, L. (2016). Disclosure of internal control information by Chinese cross listed companies. Research Report Series ISSN 2357-206X. Unitec ePress Research Report Series (1). Retrieved from http://www.unitec.ac.nz/epress/. https://pdfs.semanticscholar.org/683d/3234bdd3d3b9c52cbee786b9ced0e89ef225.pdf.

Leng, J. and Ding, Y. (2011). Internal control disclosure and corporate governance: Empirical research from Chinese listed companies. *Technology and Investment*, 2(4), pp. 286–294.

Michelon, G., Beretta, S. E., and Bozzolan, S. (2009). Disclosure on internal control systems as substitute of alternative governance mechanisms. Available at SSRN 1316323.

Owusu-Ansah, S. and Ganguli, G. (2010). Voluntary reporting on internal control systems and governance characteristics: An analysis of large US companies. *Journal of Managerial Issues*, 22(3), pp. 383–408.
Sjarief, J. and Weli, W. (2016). The internal control disclosure, the executive compensation, and the timeliness of financial reporting. *International Journal of Applied Business and Economic Research*, 14(2), pp. 991–1002.
Van de Poel, K. and Vanstraelen, A. (2011). Management reporting on internal control and accruals quality: Insights from a "comply-or-explain" internal control regime. *Auditing: A Journal of Practice & Theory*, 30(3), pp. 181–209.
Weli, W. and Sjarief, J. (2017). Analysis characteristics for internal control disclosure: Case study of companies listed in the Indonesian Stock Exchange, Global Review of Accounting and Finance, Vol. 8. No. 2. September 2017 Issue. Pp. 1–19.
Weli, W. and Sjarief, J. (2018). The effect of internal control disclosure on financial information quality and market performance distinguished by the Corporate Governance Index, International Journal of Accounting and Financial Reporting, Vol. 8, No. 1, 2018, pp. 241–260.

Facing Global Digital Revolution – Nirmala Arum Janie, Dwi Mulyaningsih & Wahyu Rachmawati (eds)
© 2020 Taylor & Francis Group, London, ISBN 978-0-367-33912-8

Rural microfinance in East Sumba: The role of local wisdom

A.D.R. Atahau & A.D. Huruta
Universitas Kristen Satya Wacana Salatiga, Salatiga, Indonesia

ABSTRACT: Microfinance has been widely used to alleviate poverty in rural areas. A comprehensive understanding of the determinant factors applicable to microfinance sustainability is required for the effective implementation of this type of finance in rural areas. This study aimed to contribute to the development of microfinance models and policy formulation in order to alleviate poverty in rural communities in East Sumba. Researchers collected data from a sample of 100 microfinance stakeholders through questionnaires distributed to community leaders, several nongovernmental organizations, and locals. The quantitative method with partial least square was applied to analyze the data. The findings showed that social, economic, and environmental factors affect microfinance practice in rural areas. Local wisdom serves as the driving force for the sustainability of microfinance. From a microfinance policy perspective, it is imperative to take into account the various factors that determine microfinance practices since a culture-based microfinance development strategy may be an alternative for local governments to consider.

Keywords: microfinance, local wisdom, sustainable development, Sumba

1 INTRODUCTION

East Sumba Regency is one of 22 districts in the East Nusa Tenggara province, which has a higher poverty rate than the Indonesian national province average. Poverty is a common characteristic of the East Sumba community that predominantly relies on agriculture with a focus on livestock. Furthermore, the average level of education is low compared to the national level. The application of basic agriculture as a lifestyle by the local population has led to entrepreneurial activities in East Sumba being dominated by migrants from other regions and provinces like Java and Bima who basically do no capital investment or profit reinvestment in East Sumba. Thus, the multiplier effect of business activities in this district is very limited. The central government was originally the sole provider of microfinance, whichis applied by way of centralized and instructive mechanisms. Therefore, the poor depended only on access to government capital. To overcome the weaknesses of this centralized government microfinance policy, East Sumba issued a communitywide economic empowerment policy known as the Integrated Rural Community Economic Empowerment for Cooperation Patterns in the early 2000s. Since 2012, the East Sumba District Planning Agency has also implemented a SMART program that targets rural communities living below the poverty line in the form of revolving funds used as business capital. However, an evaluation conducted in 2017 found that more than 50% of the funds disbursed were not repaid since the community regards money received from governing authorities as free money (Regional Development Planning Agency, 2017).

Besides the top-down approach (a government initiative), a bottom-up approach (microfinance formation initiated by society) is expected to reduce poverty through microfinance. The beliefs and values that the society adheres to and inherited from its ancestors, is based on the bottom-up approach. The approach is connected to the culture and serves to bind the communal society. This is commonly referred to as *local wisdom*, which can create positive energy that serves as the driving force for the success of the program. Soegiono, Atahau, Harijono, and Huruta (2019) show that local wisdom is based on the Marapu beliefs held by the ancestors of the Sumba community. Therefore, the development of microfinance becomes an alternative solution for rural development. However, previous research related to microfinance has not yet integrated local wisdom with microfinance development policies in poor rural areas, especially in the eastern part of Indonesia. Empirically, many factors determine the development of microfinance (Allet and Hudon, 2013). These factors can be demographic, social, or economic. Demographic factors include age, gender, and type of work (Elsayed and Paton, 2009; Jawahar and McLaughlin, 2001; Moore, 2001). Social factors include culture, local wisdom, and social capital (Bédécarrats, Baur, and Lapenu, 2012; Moore, 2001). Economic factors include poverty, entrepreneurship, and financial inclusion (Elsayed and Paton, 2009; Hulme, 2000; Kabeer, 2005; Mosley, 2001; Orlitzky, 2001; Robinson, 2002; Stanwick and Stanwick, 1998).

García-Perez, Muñoz-Torres, and Fernández-Izquierdo (2016) research the effect of four exogenous variables on the sustainability of microfinance: social, environmental, governance, and economic factors. Soegiono et al. (2019) develop an integrated local wisdom rural microfinance model that is aligned with Elkington's (1998) triple bottom line concept – namely people, profit, and the planet – also known as social, economic, and environmental aspects. This study also followed the triple bottom line concept by using social, economic, and environmental variables as determinants of microfinance sustainability. However, given the role of governance in the sustainability of microfinance as found by García-Perez et al. (2016), this study also used governance as an additional factor. In particular, this study considered local wisdom the integrated component that, with due consideration of all external variables, serves as the driving force for the sustainability of microfinance in the East Sumba rural area. This research was expected to contribute to the identification of microfinance sustainability determinants for the development of a comprehensive rural microfinance model. In addition, the implementation of this model was expected to increase the accessibility of the rural community to microfinance institutions (Beisland, Mersland, and Strøm, 2015).

2 LITERATURE REVIEW

In Indonesia, microfinance is an instrument that can be applied to achieve economic empowerment in societies like those in rural areas that experience poverty due to their economic activities like basic agriculture and low levels of formal education. The presence of bottom-up microfinance serves as the basis for managing the local financial aspirations of groups of people based on trust, norms, and rules in a network of communal partners. The network facilitates coordination and strengthens coalitions within the group. The existence of bottom-up microfinance forms an inherent part of Indonesian culture, especially communal culture. It is believed that microfinance is a means of strengthening social relations between individuals in society. Thus, microfinance has a dual economic and social function. Empirically, many factors determine the development of microfinance (Allet and Hudon, 2013). This study used economic, environmental, and social variables as determinants of microfinance sustainability with governance variables added as mediating variables.

3 RESEARCH METHODS

The population of this study comprised microfinance stakeholders in Waingapu, East Sumba. The research consisted of 100 samples obtained by using purposive sampling in two locations, namely the Tapa Walla Badi microfinance group in Mbatakapidu and the Gapoktan microfinance group in Wangga. Focus group discussions conducted in those locations followed the questionnaire distribution. Partial least square (PLS) analysis was performed.

4 RESULTS AND DISCUSSIONS

Based on the data obtained in this study, the 100 respondents have the characteristics described in this research. Most of the respondents (48%) are older than 40 years, there is an equal number of males and females (50%), and most of the respondents (77%) did not complete a senior high school degree. The respondents' low level of education may explain why the majority of them (94%) are farmers.

The model causality test is presented in the following table:

Table 1. Causality test result.

Path	Original Sample	T-statistics	P-value	Result
Governance←Economic	0.416	2.726	0.007	Accepted
Microfinance←Economic	0.535	2.030	0.043	Accepted
Governance←Environment	0.271	1.701	0.090	Accepted
Microfinance←Environment	0.094	0.441	0.659	Rejected
Microfinance←Governance	0.058	0.190	0.849	Rejected
Governance←Social	0.328	2.228	0.026	Accepted
Microfinance←Social	-0.017	0.063	0.949	Rejected

Source: The output of primary data using SmartPLS-v.3, 2019

Based on these findings, the economic variable has a positive influence on the governance and sustainability of microfinance, the environmental factor has a positive influence on the governance of microfinance, and the social factor has a positive influence on the governance of microfinance. On the other hand, the direct effect of the environmental and social factors on microfinance sustainability is rejected. The results of this study provided empirical evidence that economic, social, and environmental factors are the determinants of the governance of microfinance, and the sustainability of microfinance is determined by the economic factor. This provides evidence that economic factors are the primary factors affecting microfinance sustainability. This is in line with the findings of Elsayed and Paton (2009), Hulme (2000), Kabeer (2005), Mosley (2001), Orlitzky (2001), Robinson (2002), and Stanwick and Stanwick (1998), where economic factors such as income, assets, and networks show the ability of *arisan* group members to maintain their contribution to the groups as a form of microfinance. Interesting, all factors (economic, social, and environmental) that consider local wisdom affect the governance of microfinance institutions. This is supported by the focus group discussion conducted with the sample groups, since the key discussion information emphasized the importance of governance in managing microfinance institutions. The results show that local wisdom, embedded in each factor, plays a significant role in the governance of microfinance institutions. The trust that bind the members and managers (*pawandagu* and *panjulurungu*), and the personal approach and empathy in governance (*mandara*), are some of the underlying local wisdom components that are inherent in each factor that affects governance. However, the results do not provide sufficient evidence to support a direct relationship between governance and microfinance sustainability.

5 CONCLUSION

The results of this study provided empirical evidence that economic, social, and environmental factors are the determinants of the governance of microfinance, and microfinance sustainability is determined by economic factors. Local wisdom serves as the supporting foundation for the effect of these factors on the sustainability of rural microfinance. In addition, this study showed that the exogenous governance factor – which, according to the García-Perez model of microfinance, is a determinant of microfinance sustainability – depends on economic, social, and environmental factors. From a microfinance policy perspective, it is imperative to take into account the various factors that determine microfinance practices since a culture-based microfinance development strategy may be an alternative for local governments to consider.

REFERENCES

Allet, M. and Hudon, M. (2013). Green microfinance: Characteristics of microfinance institutions involved in environmental management. Journal of Business Ethics, 126(3), pp. 395–414. Available at: https://doi.org/10.1007/s10551-013-1942-5.

Bédécarrats, F., Baur, S., and Lapenu, C. (2012). Combining social and financial performance: A paradox? Enterprise Development and Microfinance, 23(3),pp. 241–258. Available at: https://doi.org/10.3362/1755-1986.2012.024.

Beisland, L. A., Mersland, R., and Strøm, R. Ø. (2015). Audit quality and corporate governance: Evidence from the microfinance industry. International Journal of Auditing, 19(3), pp. 218–237. doi: 10.1111/ijau.12041.

Elkington, J. (1998). *Cannibals with Forks: The Triple Bottom Line of 21st Century Business*. Gabriola Island, BC: New Society Publishers.

Elsayed, K. and Paton, D. (2009). The impact of financial performance on environmental policy: Does firm life cycle matter? Business Strategy and the Environment, 18(6),pp. 397–413. Available at: https://doi.org/10.1002/bse.608.

García-Pérez, I., Muñoz-Torres, M. J., and Fernández-Izquierdo, M. Á. (2016). Microfinance literature: A sustainability level perspective survey. Journal of Cleaner Production, 142, pp. 3382–3395. doi: 10.1016/j.jclepro.2016.10.128.

Hulme, D. (2000). Impact assessment methodologies for microfinance: Theory, experience and better practice. *World Development*, 28(1), pp. 79–98. doi: 10.1016/S0305-750X(99)00119-9.

Jawahar, I. M. and McLaughlin, G. L. (2001). Toward a descriptive stakeholder theory: An organizational life cycle approach. Academy of Management Review, 26(3),pp. 397–414. Available at: www.jstor.org/stable/259184.

Kabeer, N. (2005). Is microfinance a "magic bullet" for women's empowerment? Analysis of findings from South Asia. Economic and Political Weekly, 40(44/45),pp. 4709–4718. Available at: www.microfinancegateway.org/sites/default/files/mfg-en-paper-is-microfinance-a-magic-bullet-for-womens-empowerment-analysis-of-findings-from-south-asia-oct-2005_0.pdf.

Moore, G. (2001). Corporate social and financial performance: An investigation in the U.K. supermarket industry. Journal of Business Ethics, 34(3/4),pp. 299–315. Available at: https://link.springer.com/article/10.1023/A:1012537016969.

Mosley, P. (2001). Microfinance and poverty in Bolivia. Journal of Development Studies, 37(4), pp. 101–132. doi: 10.1080/00220380412331322061.

Orlitzky, M. (2001). Does firm size confound the relationship between corporate social performance and firm financial performance? Journal of Business Ethics, 33(2),pp. 167–180. Available at: https://link.springer.com/article/10.1023/A:1017516826427.

Regional Development Planning Agency. (2017). Laporan Akhir Evaluasi Program SMART. East Sumba.

Robinson, M. S. (2002). *The Microfinance Revolution: Lessons from Indonesia*. Washington, DC: The World Bank and Open Society Institute. Available at: http://hdl.handle.net/10986/14254.

Soegiono, L. et al. (2019). Local wisdom in rural microfinance: A descriptive study of the villagers of East Sumba. Entrepreneurship and Sustainability Issues, 6(3), pp. 1485–1496. doi: 10.9770/jesi.2019.6.3(30).

Stanwick, P. A. and Stanwick, S. D. (1998). The social and performance relationship between corporate organizational size, financial performance, and environmental performance: An empirical examination. Journal of Business Ethics, 17, pp. 195–204. doi: 10.1023/A:1005784421547.

Facing Global Digital Revolution – Nirmala Arum Janie, Dwi Mulyaningsih & Wahyu Rachmawati (eds)
© 2020 Taylor & Francis Group, London, ISBN 978-0-367-33912-8

Capital market reactions to commercial aircraft accidents

Dyah Nirmala Arum Janie, Tri Agustin Fatmasari, Yulianti Yulianti & Abdul Karim
Universitas Semarang, Semarang, Indonesia

ABSTRACT: Noneconomic events are one of the main factors that might affect capital markets. This Indonesia Stock Exchange (IDX) research aimed to test activities involving aviation accidents. We analyzed stock price and trade volume after several commercial aviation accidents in 2014–2018: Air Asia (December 28, 2014), Trigana (August 16, 2015), and Lion Air (October 29, 2018). We adopted quantitative methods with secondary data of daily stock price summary. The population comprised aviation companies listed on the IDX, and the period observed was three days, five days, and seven days before and after the accidents. Paired sample t-tests were conducted for normally distributed data, along with the Wilcoxon signed rank test and vice versa. The average abnormal return of five days before and after the Trigana accident represented a significant difference. It meant that the event has information content that can cause capital markets to react. Before and after the Air Asia and Lion Air accidents, the abnormal return differed insignificantly. The results for the Air Asia, Trigana, and Lion Air accidents also had insignificant differences.

Keywords: abnormal return, study event, trading volume activity, aviation accidents

1 INTRODUCTION

In the past five years, several airplane accidents have occurred in Indonesia. On December 28, 2014, Air Asia flight QZ8501 (carrying 156 passengers) crashed in the waters near Pangkalan Bun, Central Kalimantan, shortly after taking off from Juanda Airport, Surabaya. On August 16, 2015, a Trigana Air aircraft (carrying 54 passengers) with registration number PK YRN lost contact during its Jayapura–Oksibil flight; public information claimed the plane crashed into the mountains on Okbape. Finally, on October 29, 2018, a Lion Air JT 610 aircraft (carrying 189 passengers), on the Soekar–Hatta route from Tangerang to Pangkalpinang, fell in Tanjungpakis, Karawang, after flying from Soekarno Hatta International Airport, at 06.20 WIB. Based on data collected, this incident was the twentieth aviation accident experienced by Lion Air since 2002.

Events that occur around us are often analyzed in order to evaluate the capital market response, given how strongly these events affect investors' decisions to invest in related companies. In this study, the author used the indicator of the difference in abnormal return and trading volume activity before and after the events, because in general the information content, market reaction, and level of market efficiency are measured and tested using these variables, and have been needed in previous studies.

2 LITERATURE REVIEW

Research with event studies aims to evaluate the market reaction to an event whose information has been made public. The application of economic and noneconomic events has often been made well in previous studies. However, researchers have focused more on event research using study methods that concentrate on economic events.

Ho, Qiu, and Tang (2013) describe plane crashes in the period 1950–2009 using a sample of 113 air accidents and 1,199 flights that did not experience an accident from ASN Aviation Safety. This study found no significant abnormal returns consistently related to the events.

Rahmawati (2016) tests the average abnormal return on LQ45 companies listed on the Indonesia Stock Exchange (IDX) in the period December 2015–January 2016, using the Sarinah Plaza bombing. The results of the study found no difference in average abnormal return on LQ45 stocks both before and after the Sarinah Plaza bombing.

Widyasari, Suffa, and Praswati (2017) examine the impact of the 2016 tax amnesty policy on companies at 10 days before and after the event, namely September 16–October 14, 2016. The samples taken in this study were the shares of 26 companies listed on the IDX. The results of the study showed no positive abnormal return, which indicated the reaction of market participants to the decline in stock wealth.

Purba and Handayani (2017) analyze the stock price response of LQ45 companies listed on the IDX from February to July 2017. Their study uses political events related to the 2017 second round of the DKI Jakarta regional elections. The results of testing the significance of abnormal returns found a significant adverse market reaction; besides that, there is no average abnormal returns and average trading volume activity between the period before and after the event.

Nisa (2017) examines the impact of the 2014 Air Asia aircraft accident on stock investments in the Malaysian Stock Exchange. Samples taken included 15 travel and leisure industry companies. The results of the analysis show abnormal return and trading volume activity around the day of the Air Asia crash, which is five days before and five days after the event, a statistically significant difference. A difference occurred in abnormal stock returns in the travel and leisure industry between before and after the Air Asia crash, but no significant difference in trading volume.

Gumanti, Savitri, Nisa, and Utami (2018) test abnormal return and trading volume activity on shares of travel and leisure industry companies listed on the Kuala Lumpur Stock Exchange with the event of the Air Asia crash, using 15 samples of shares of companies actively traded in the period five days before and five days after the event. The results of this study indicate no abnormal return concerning these events. However, differences appeared in abnormal returns between before and after the Air Asia crash, in contrast to the trading volume, which saw no significant difference.

Study events analyze market reactions to the content of a public announcement or notification. If the announcement contains meaningful information, the market will respond. The average abnormal return every day can be different or not at all. It depends on how investors react to the news and how quickly the market reacts to absorb information in order to reach a new equilibrium state. Information related to an asset is investigated so that the effect it produces is a change to a new equilibrium price. This equilibrium price remains stable until more new information converts it back to the new equilibrium price. Nisa (2017), in his research related to the impact of the Air Asia aircraft accident on stock returns in the Malaysian Stock Exchange, finds that the average abnormal return on travel and leisure industry shares before the event was different from abnormal returns after the event.

Based on the description, we hypothesized that:

H1: Differences appear in the average significant abnormal return before and after commercial aircraft accidents in the aviation sector shares listed on the IDX.

Trading volume activity (TVA) is an instrument that can be used to observe capital market action on information through parameters of changes in TVA in the capital market (Suryawijaya and Setiawan, 1998, cited in Hidayati, Maslichah, and Junaidi, 2017). Commercial aircraft accidents are unexpected events seen as having information content, so the market is expected to respond. Based on this description, we hypothesized that investors respond to commercial aircraft accidents in the presence of differences in the average TVA in the period before and after commercial aircraft accidents.

H2: A significant difference emerges in the average TVA before and after commercial aircraft accidents in the aviation sector shares listed on the IDX.

3 METHOD

This research used quantitative methods. The population of this study comprised the airline transportation sector companies listed on the IDX. The observation period was every three, five, and seven days in the period before and after the event, and the data used were secondary data in the form of a daily summary of shares. Tests carried out with different test methods paired sample t-tests with normally distributed data, and the Wilcoxon signed rank test was used for data that were distributed unevenly.

4 RESULTS AND DISCUSSIONS

Hypothesis 1, which was tested in the five days before and after the Trigana Air accident on August 16, 2015, showed a significant difference in the average abnormal return. We can see from the results of the study where the value (Sig. 2-tailed) of the cumulative average abnormal return for five days before and after the Trigana event is smaller than the level of significance (α) 5%, which is 0.001 < 0.05, so H1 was accepted. The difference in abnormal returns in the five days before and after the Trigana event shows that the market of investors in the IDX was affected by the accident. We can see this from the closing stock price of PT. Garuda Indonesia (Persero) Tbk, which attracted the attention of investors in investing in PT. Garuda Indonesia (Persero) Tbk due to the Trigana accident.

We tested H1 for three days before and seven days after the Trigana accident, then we tested it for three, five, and seven days before and after the Air Asia accident on December 28, 2014. We later tested for three, five, and seven days before and after the Lion Air accident on October 29, 2018, which showed no significant difference in the average abnormal return. The result shows the value (Sig. 2-tailed) of the cumulative average abnormal return is higher than the level of significance (α) 5%, so we can only reject H1.

The absence of a difference in abnormal returns in all the periods before and after the commercial airplane crashes shows that the investor market in the IDX was not affected by the accidents. We can see this from the closing stock price of PT. Garuda Indonesia (Persero) Tbk and PT. Air Asia Indonesia Tbk, which has not changed due to the event. In other words, the presence or absence of an aircraft accident during this period did not affect the reaction of investors in trading shares on the IDX.

We tested H2 for three, five, and seven days before and after the Air Asia accident. Then we tested it for three, five, and seven days before and after the Trigana accident. Later, we tested it for three, five, and seven days before and after the Lion Air accident. We found no significant difference in the average TVA. The results show the value (Sig. 2-tailed) of cumulative average abnormal return is higher than the level of significance (α) 5%. Thus H2 is also rejected, which means there is no difference in the average TVA between before and after the commercial airplane accidents. The trading volume of PT. Garuda Indonesia (Persero) Tbk and PT. Air Asia Indonesia Tbk experienced a random increase and decrease. this means that the events of commercial aircraft accidents do not illicit a positive response on the stock market.

Hypothesis testing of the company's abnormal return variable shows the following results. The periods of three, five, and seven days before and after the Air Asia accident obtained a p value > 0.05; this means the value of the abnormal return of PT. Garuda Indonesia (Persero) Tbk had no significant difference between before and after an accident. We can conclude that our three hypotheses are not accepted.

The period of three days before and seven days after the Trigana accident obtained a p value > 0.05; this means the value of the abnormal return of PT. Garuda Indonesia (Persero) Tbk saw no significant difference between before and after an accident. We can conclude that our two hypotheses are not accepted. In the period of five days before and after, however, the p-value < 0.05; this indicates a significant difference between before and after the accident. So it can be concluded that this hypothesis is accepted.

5 CONCLUSION

This study had many limitations, including that not all the hypotheses it analyzed showed differences in abnormal return and TVA. Therefore, further research should, in addition to performing technical analysis, pay attention to fundamental analysis in investment decision-making that can benefit from the fundamental analysis side. Then, we also recommend not only testing the information content of an event but also continuing with testing market efficiency. Future research can also consider the corporate sector used, so that the research sample is not limited only to aviation transportation sector companies.

REFERENCES

Anonim, A. (2018, October 31). Retrieved December 1, 2018, from www.pikiran-rakyat.com: www.pikiran-rakyat.com/nasional/2018/10/31/daftar-panjang-kecelakaan-pesawat-di-indonesia-dalam-14-tahun-terakhir-432453.

Gumanti, T. A., Savitri, E., Nisa, N. W., and Utami, E. S. (2018). Event study on the crash of the Air Asia plane: A study of travel and leisure companies listed on the Malaysian Stock Market. *Jurnal Akuntansi dan Keuangan*, 20(1), pp. 20–26.

Ho, J. C., Qiu, M., and Tang, X. (2013). Do airlines always suffer from crashes? *Economic Letters*, 118, pp. 113–117.

Nisa, N. W. (2017). Peristiwa jatuhnya Air Asia dan Gejolak Pasar Malaysia. *Prosiding Seminar Nasional dan Call for Paper Ekonomi dan Bisnis*, pp. 372–382.

Rahmawati, I. Y. (2016). Reaksi pasar modal dari dampak peristiwa bom Plaza Sarinah terhadap abnormal return perusahaan LQ 45 yang terdaftar di BEI. *Riset Akuntansi dan Keuangan Indonesia*, 1(2), pp. 126–133.

Widyasari, T. N., Suffa, I. F., and Praswati, A. N. (2017). Analisis reaksi pasar modal atas peristiwa kebijakan amnesti pajak 2016. *Jurnal Administrasi Bisnis*, 6(2), pp. 137–146.

Purba, F., & Handayani, S. R. (2017). Analisis Perbedaan Reaksi Pasar Modal Indonesia Sebelum Dan Sesudah Peristiwa Non Ekonomi (Studi pada Peristiwa Politik Pilkada DKI Jakarta 2017 Putaran Kedua). *Jurnal Administrasi Bisnis (JAB)*, 51(1),115–123.

Hidayati, N., Maslichah, M., & Junaidi, J. (2018). REAKSI ABNORMAL RETURN DAN TRADING VOLUME ACTIVITY TERHADAP RAMADHAN EFFECT (Studi pada Perusahaan Food and Beverages yang Terdaftar di Bursa Efek Indonesia Periode 2015-2017). *Jurnal Ilmiah Riset Akuntansi*, 7(03).

Facing Global Digital Revolution – Nirmala Arum Janie, Dwi Mulyaningsih & Wahyu Rachmawati (eds)
© 2020 Taylor & Francis Group, London, ISBN 978-0-367-33912-8

Determinants of construction, property, and real estate companies' stock prices in Indonesia

Dyah Nirmala Arum Janie, Laeli Tika Mardani, Dian Indriana Tri Lestari & Nirsetyo Wahdi
Universitas Semarang, Semarang, Indonesia

ABSTRACT: This study aimed to determine the effect of leverage, liquidity, profitability, and rupiah exchange rates on stock prices of Indonesian construction, property, and real estate companies listed on the Indonesia Stock Exchange from 2014 to 2018. Independent variables included profitability, which comprises net profit margin (NPM), return on equity (ROE), return on assets (ROA), and earnings per share (EPS). This study examined the variables that have the most significant influence on stock prices. This research employed a documentary method with a quantitative approach. Descriptive research and verificative exploration were used. The research data consisted of time series and cross-sectional data. They depicted a situation at a particular time. Purposive sampling obtained a sample of 45 companies. The data used were secondary data, where the leverage, liquidity, profitability, and rupiah exchange rates served as the dependent variable. The partial least square (PLS) method was utilized.

Keywords: leverage, liquidity, profitability, IDR exchange rates, stock prices

1 INTRODUCTION

The current population growth has meant that community needs such as homes, offices, shopping centers, entertainment, and schools have increased. As a result, the construction, property, and real estate sectors are also experiencing growth, which presents a significant market opportunity for these sectors in Indonesia. Information like this is crucial and decisive for potential investors, who might later respond by buying shares in companies that operate in these industries (Eka and Suratno, 2015).

Companies engaged in these sectors have been known to have volatile, persistent, and complex characteristics where prices are one of the most important financial indicators. It is appropriate if investors believe that they will get a big profit if they invest their capital in the construction, property, and real estate sectors because the supply of land is permanent, but demand is always high, as is the current population growth.

2 LITERATURE REVIEW

Factors that can influence stock prices, according to Tenriola and Akramunnas (2017), include fundamental and technical analyses, where vital information can be seen about a company's financial performance. Fundamental information is internal to a company, while technical information comes from external sources influenced by macro conditions. Most investors make decisions according to macro conditions, where when these conditions experience a decline, the value of the stock will also follow a downturn. If economic conditions improve or can be said to be healthy, they can also have an impact on rising stock prices. In this study, the external variables impacting current conditions included the variable exchange rate.

Analysis of corporate financial performance is critical before deciding to invest in a company. The condition of a company, which has a close relationship with financial performance, can positively affect stock prices. Financial statements provide a real assessment of how a company performs. Usually, financial statements of companies that have gone public are published in the financial report on the Indonesia Stock Exchange (IDX).

The financial ratio compares the number for each posted financial figure with the number of other posted figures, according to Fahmi and Irham (2014). Financial ratios consist of liquidity ratios, leverage, profitability ratios, and activity ratios. Eka and Suratno (2015) reveal that if the four ratios result in a good value and show a stable condition, then stock prices may be high. Investors also need to consider the current ratio, where the liquidity ratio is advantageous, to measure a company's performance in terms of paying off short-term debt.

An attractive company enjoys high profit margins in generating cash profits for its owners, so a net profit margin (NPM) ratio is needed to evaluate company performance. Likewise, earnings per share (EPS) is crucial information in describing the outlook for a company's earnings in the future. Other profitability ratios, such as return on assets (ROA), are useful for measuring the productivity of all company funds from the assets the company has. Return on equity (ROE) is useful to measure a company's ability to glean profits from investments.

The debt-to-equity ratio (DER) is one of the leverage ratios that is useful for measuring the ratio of debt to equity; if the ratio is high, the probability of using debt is higher than the equity. Other domestic factors that can affect shares is the rupiah exchange rate against foreign currencies. The rupiah exchange rate always fluctuates from year to year. It makes stock prices unstable, which results in investors not being interested in investing in the capital market and trading on the stock exchange, weakening and influencing stock prices.

The rise of economic activity in the construction, property, and real estate sectors is an indicator used in analyses of the economic health of a country. Outstanding financial performance will attract investors, and it promotes competition between companies that later can be taken into consideration and can provide consistent benefits. Investors will see the condition of the company first, followed by sound financial performance.

3 METHOD

The research data were in the form of time-series and cross-sectional data that reflect a picture of a situation at a specific time. Purposive sampling obtained a sample of 45 companies. The data used were secondary data, where leverage, liquidity, profitability, and rupiah exchange rates served as independent variables and stock prices comprised the dependent variable. The statistical method used was partial least square (PLS). The results of this study proved that profitability ratios have a significant effect on stock prices, while leverage, liquidity, and rupiah exchange rates have no significant effect on stock prices.

4 RESULTS

Leverage variables measure the ratio of debt to equity. In this study, the leverage variable used was the measurement of the DER ratio, where the DER's influence on stock prices showed a coefficient of −0.082 with a significance level of $0.107 > 0.05$. This means that an increase in DER of one unit will be followed by a decrease in stock price of 0.0082. This research is consistent with that done by Nita and Silviana (2016), which finds that leverage ratios do not have a significant effect on stock prices.

We can interpret the current ratio as a company's ability to pay off its short-term obligations by using current assets owned (Imelda, Wibowo, & Diarsyad 2018). Based on testing using WarpPLS, the current ratio variable does not have a significant effect on stock prices, with a significance value of $0.471 > 0.05$ and a coefficient value of −0.005, which means that an increase in the current ratio of one unit is followed by a decrease in the share price of −0.005. Investors therefore do not see the current ratio as a factor in their decision to buy shares, which is in line with research conducted by Pande and Nyoman (2018) and Nita and Silviana (2016), which finds that liquidity ratios do not significantly influence stock prices.

Profitability can be interpreted as a net result of a series of policies and decisions that allow a company to survive, where a company must be in a favorable condition. Without any benefits, it will be difficult for a company to attract outside capital. The variable of profitability had four indicators with a loading factor value that is per the rule of thumb, which was then tested again with confirmatory analysis with results above 0.40 with a P value less than 0.05, which means that this indicator can explain the profitability variable well. The indicator with the highest value for profitability was ROE; the higher the ratio, the better the company in measuring the ability of its own capital to generate profits for all shareholders (Jestry, Paulina, and Dedy, 2017).

If a company can be said to be good, of course its stock price also rises, which is supported by previous research, namely Jelie, Parengkuan, and Jeffry (2017), Imelda, Wibowo, & Diarsyad (2018), and Luvy (2014). The lowest indicator was EPS with a loading factor of 0.513; this means that investors only see the ROE value before investing in a particular company, because it is very reasonable for investors to be attracted to a stock that will return a significant level of profit from their capital.

The rupiah exchange rate is one of the external variables that affected stock prices in this study. The rupiah exchange rate can affect Indonesia's domestic industry. If the rupiah exchange rate appreciates at the exchange rate of a country's currency, we can say that the share price will increase. In processing data from WarpPLS, we can conclude that the rupiah exchange rate coefficient is 0.063, which means that when an increase in the rupiah exchange rate of one unit can be said to be followed by an increase in the stock price of 0.063, the exchange rate has a positive relationship to stock prices. This corroborates Vitra and Saparila (2018). However, in this study, the rupiah exchange rate had a significance value of 0.172, which means that the rupiah exchange rate is not significant for stock prices.

5 CONCLUSION

The research sample comprised only 45 companies, because many companies have not published financial statements for the 2018 period, and many companies are also incomplete in issuing financial statements in a row. Partial least square testing in this study had a weakness. We could measure some variables by only one indicator (leverage ratio, liquidity ratio, and the rupiah exchange rate). We consider this a weakness because when measuring the model/outer model to find out whether indicators could measure the tested variables, using one indicator did not meeting the loading factor requirements (> 0.70), so the variable was not measurable. However, using PLS in this study obtained the right indicators for certain variables, for example, the measurement of profitability ratios found four indicators: namely ROA, ROE, NPM, and EPS.

Future researchers need to choose other industries in order to obtain more accurate results and to heighten the empirical testing of the effects of leverage, liquidity, profitability, and exchange rates on stock prices. Researchers could add an observation period. The longer the observation time interval, the higher the opportunity to obtain information about reliable variables in order to make accurate assessments. We also could add independent variables obtained from external conditions such as inflation, interest rates, and the money supply to stock prices, because it is possible that these factors have a stronger influence on stock prices (Imelda, Wibowo, & Diarsyad 2018).

REFERENCES

Eka, B. and Suratno, S. (2015). Return on equity, debt to equity ratio, price earning ratio, assets growth, inflasi dan return saham perusahaan property dan real estate. *Jurnal Riset Akuntansi dan Perpajakan*, 2, pp. 153–166.

Fahmi, I. (2014). *Pengantar Manajemen Keuangan*. Bandung: Alfabeta.

Imelda, & Wibowo, Agus & Diarsyad, Muhammad. (2018). Pengaruh current ratio, cash ratio, return on equity dan return on assets terhadap harga saham perusahaan property dan real estate yang terdaftar di BEI. *Jurnal Riset Akuntasi Keuangan*, pp. 102–124.

Jelie, D., Parengkuan, and Jeffry. (2017). Pengaruh struktur modal, ukuran perusahaan dan profitabilitas terhadap harga saham pada perusahaan industri sektor makanan dan minuman yang terdaftar di Bursa Efek Indonesia. *Jurnal EMBA*, 5, pp. 3385–3394.

Jestry, J., Paulina, and Dedy. (2017). Analisis pengaruh profitabilitas terhadap harga saham pada perusahaan yang terdaftar di Bursa Efek Indonesia Di LQ 4., *Jurnal EMBA*, 5, pp. 753–761.

Luvy, N. (2014). Pengaruh kinerja keuangan terhadap return saham perusahaan property dan real estate. *Jurnal Bisnis dan Manajemen*, 6, pp. 89–97.

Nita, F., and Silviana. (2016). Pengaruh likuiditas, solvabilitas, profitabilitas, rasio pasar dan ukuran perusahaan terhadap harga saham (sudi pada perusahaan subsektor perkebunan yang terdaftar di Bursa Efek Indonesia tahun 2010–2014). *Jurnal Riset Akuntansi dan Keuangan*, 4, pp. 102–124.

Pande, W. and Nyoman. (2018). Pengaruh EPS, PER, CR dan ROE terhadap harga saham di Bursa Efek Indonesia. *E-Journal Manajemen Unud*, 7, pp. 2106–2133.

Tenriola, S. and Akramunnas. (2017). Pengaruh faktor fundamental dan teknikal terhadap harga saham industri perhotelan yang terdaftar di Bursa Efek Indonesia. *Jurnal Ekonomi, Keuangan dan Perbankan Syariah*, 1, pp. 116–131.

Vitra, I. and Saparila. (2018). Pengaruh tingkat inflasi, nilai tukar rupiah, dan tingkat suku bunga domestik terhadap indeks harga saham syariah Indonesia. *Jurnal Adsministrasi Bisnis*, 60, pp. 119–128.

Facing Global Digital Revolution – Nirmala Arum Janie, Dwi Mulyaningsih & Wahyu Rachmawati (eds)
 ISBN 978-0-367-33912-8

Exploration of social responsibility implementation model in small and medium micro enterprises

Indarto & Aprih Santoso
Universitas Semarang, Semarang, Indonesia

P. Chatarina Yekti
Soegijapranata Catholic University, Semarang, Indonesia

ABSTRACT: The purpose of this study was to explore the model of the implementation of corporate social responsibility by MSMEs. The focus of the research was the activities carried out by MSMEs in implementing CSR. The method of data collection in this study was observation and in-depth interviews with MSME actors. The respondents of this study were the actor of MSMEs in Semarang who has run the business for at least 3 years. The exploratory studies were chosen to find patterns of implementation of CSR carried out by MSMEs towards employees, consumers, society and the environment. The model for implementing CSR by MSMEs found in the implementation of CSR that is integrated with routine business activities. The implementation of CSR by MSMEs is activities aimed at serving stakeholders as well as possible based on morality, ethics, and empathy for society and the environment. We propose that the model be used as a guide for MSMEs in implementing CSR principles.

Keywords: model, implementation, CSR, MSMEs, stakeholders, ethics

1 INTRODUCTION

The huge role of MSMEs in the local and national economy is undeniable. Based on data from Statistics Indonesia, contribution's SMEs to the GDP in 2017 reached 60.34 percent. Absorption of labor in the MSME sector in the same period reached 97.22 percent in the same period. In addition to contributing economically, MSMEs also contribute socially. The social impact of the presence of MSMEs is to reduce unemployment, accelerate poverty alleviation and encourage community productivity and creativity. MSMEs usually move towards processing the closest resources around them so that MSMEs also play a role in maintaining local traditions and wisdom.

The role of MSMEs that are so large economically and socially must be supported so that they can provide greater benefits to the wider community. If MSMEs apply the concepts of corporate social responsibility (CSR) to all their stakeholders in running a business, this will greatly support the existence of MSMEs. The implementation of CSR by MSMEs should also receive great attention, such as the implementation of CSR by large businesses. So far, research on CSR has focused on large companies because it is considered a large company that is capable of carrying out CSR.

Research by Larrán Jorge *et al.*, (2016) concluded that MSMEs were also able to carry out CSR. Research on the implementation of CSR in Malaysia conducted by Atan and Halim (2012) concluded that there was high motivation for MSMEs to implement CSR. CSR implementation can prevent conflicts of interest between stakeholders; improve reputation and competitive advantage of MSMEs. While the research from (Coppa and Sriramesh, 2013; Frynas, 2016; Turyakira, 2014; Witjes et al., 2015)) confirms that companies on small and medium scale will gain future benefits if they implement their social responsibility to their stakeholders.

The contribution of MSMEs to society and the economy will be increasingly meaningful if MSMEs also carry out social responsibility towards their stakeholders (Pastrana and Sriramesh, 2014). Stakeholders of MSMEs are usually communities around these MSMEs. Internal stakeholders of MSMEs are MSME workers themselves. Labor intensive MSMEs will have a very significant impact on the welfare of workers and society. The closeness of the owner and its stakeholders makes it easier for MSMEs to integrate CSR values than large businesses. While external MSME stakeholders are communities around these MSMEs.

Research by Andy and Mustapha (2013) concluded that the implementation of CSR by MSMEs depends on the character of the MSME actors themselves. The character referred to here is a character that is influenced by religion, culture and spiritual values it has. These religious, cultural and spiritual values encourage MSME actors to conduct business honestly and be responsible for all stakeholders. According to the results of research by Copaa and Sriramesh (2012) and Atan and Halim (2012), the implementation of CSR by MSMEs was motivated more by the values and beliefs held by MSME owners.

Research to explore models of CSR implementation that can be applied by MSMEs is very important. The model for implementing CSR by MSMEs is needed to become a guide for every MSME in carrying out CSR principles so that MSMEs can enjoy sustainable competitive advantages.

2 RESEARCH METHODS

The sample of this study was MSME actors in Semarang City. Sampling using purposive random sampling technique. The criteria for MSMEs that are sampled are MSMEs that produce products (not services) with a minimum duration of 3 years. The 3-year business period is assumed that MSMEs have been sufficiently able to survive in competition and have had quite intense attachments with their stakeholders. The numbers of samples for this study were 30 MSMEs that were spread in the Semarang City area. The data collection technique is by observation and in-depth interviews with the questionnaire as a guide. Observation of the sample of MSME actors was conducted to find out the implementation of CSR by MSMEs, while in-depth interviews were conducted to gather information about the motivation and experience of MSMEs in implementing CSR.

3 RESULTS AND DISCUSSION

The results of this exploratory study found a model for implementing CSR that could be done at the level of MSMEs. Model for implementing CSR in MSME scale found is a model of the implementation of CSR principles that are integrated in daily business operations. In implementing CSR, MSMEs do not specifically allocate budget and time. MSMEs implementing CSR by running a business based on ethics, morals and empathy for all stakeholders. Model the implementation of MSME scale CSR is presented as follows:

Table 1 shows the model of CSR implementation by MSMEs towards employees, consumers, society and the surrounding environment. The implementation of CSR by MSMEs is realized in activities that are integrated with the company's operations related to its stakeholders. The activities of implementing CSR by MSMEs on employees are activities related to the development of employee capabilities, equal opportunity, participation in business activities, harmonious relations between the owner and employees and their families. Activities to implement CSR by MSMEs towards the community and environment are social and environmental activities in the form of community empowerment and concern for the surrounding environment. CSR activities by MSMEs that are oriented towards the surrounding environment focus more on handling waste from their production activities.

Table 1. Model of CSR implementation by MSMEs.

CSR to employees	CSR to Consumers	CSR to Society	CSR to Environment
Conditioning a safe and comfortable work environment for employees	Ensure that the products produced are safe and healthy for consumers	Using labor from the surrounding community	Using environmentally friendly packaging
Providing adequate compensation and welfare for employees	Ensure products sold do not contain substances or elements that are harmful to consumers	Always contribute to the activities of the surrounding community	Processing waste with regard to environmental safety and health
Give bonuses to diligent/accomplished employees	Always include expiration dates, usage methods or presentation suggestions	Make use of resources or raw materials in the surrounding environment	Initiating the movement to care about the natural environment around the business
Give holiday allowances	List ingredients in the product packaging	Contributing to the construction of public facilities for local residents	Guarantee the production process does not cause interference with the surrounding natural environment
Give employees the opportunity to submit suggestions and input	Using packaging that guarantees product quality	Provide training or coaching for local residents	Strive for energy-efficient production processes
Give medical assistance when an employee is sick or when he experiences an accident	Providing services for consumers to submit complaints and suggestions	Interact well with local residents	Wise in using water and electricity
Organizing shared refreshing opportunities for employees	Willing to be responsible if the product turns out to harm consumers	Engaging in solving problems in the local environment	Minimizing pollution in the production process
Provide opportunities for employees to improve their skills and knowledge	Honest in measure	Build networks with local governments to assist in community empowerment	Actively involved in the Reduce, Reuse and Recycle Programs
Maintain harmonious communication between owners and employees	Set a reasonable price	Obey and respect every rule and norm that applies in the community	educates and campaigns for environmental love movements
Treat employees as work partners not helpers	Fair applies to consumers	Not greedy in pursuit of profit	Using materials that are environmentally friendly
Give holidays for employees	Have a permit and certificate that guarantees product quality and safety	Share with the powerless community	Using environmentally friendly technology

Source: primary data, 2018

4 CONCLUSION

The model for implementing CSR in MSME scale is the implementation of CSR that is integrated with the MSME business operations. CSR implementation model by MSMEs is to run the business based on ethics, morals and empathy for people and the environment.

Implementation of CSR by MSMEs is not always planned and budgeted, but is carried out at all times when interacting with stakeholders. Implementation at the MSME level is legitimized with morality and ethics. It is the moral and ethics that drive MSMEs to implement CSR in accordance with their respective abilities.

REFERENCES

Andy, L. and Mustapha, M. (2013) 'CSR in small and medium enterprises: evidence from Malaysia', in *2nd International Conference on Management, Economics and Finance (2nd ICMEF 2013) Proceeding. Sabah.*

Atan, R. and Halim, N. A. A. (2012) 'The perception of Muslim consumers towards corporate social responsibility', *PERCEPTION*, 4(2).

Coppa, M. and Sriramesh, K. (2013) 'Corporate social responsibility among SMEs in Italy', *Public Relations Review*. Elsevier, 39(1), pp. 30–39.

Larrán Jorge, M. *et al.* (2016) 'Development of corporate social responsibility in small and medium-sized enterprises and its nexus with quality management', *Cogent Business & Management*. Taylor & Francis, 3(1), p. 1228569.

Pastrana, N. A. and Sriramesh, K. (2014) 'Corporate social responsibility: Perceptions and practices among SMEs in Colombia', *Public Relations Review*. Elsevier, 40(1), pp. 14–24.

Turyakira, P., Venter, E. and Smith, E. (2014) 'The impact of corporate social responsibility factors on the competitiveness of small and medium-sized enterprises', *South African Journal of Economic and Management Sciences, 17*(2), pp.157–172.

Witjes, S., Vermeulen, W.J. and Cramer, J.M. (2017) 'Exploring corporate sustainability integration into business activities. Experiences from 18 small and medium sized enterprises in the Netherlands', *Journal of cleaner production, 153*, pp.528–538.

Facing Global Digital Revolution – Nirmala Arum Janie, Dwi Mulyaningsih & Wahyu Rachmawati (eds)
© 2020 Taylor & Francis Group, London, ISBN 978-0-367-33912-8

Ethical business practice, corporate social responsibility and competitive advantage

Indarto
Universitas Semarang, Semarang, Indonesia

P. Chatarina Yekti
Soegijapranata Catholic University, Semarang, Indonesia

Wyati Saddewisasi
Universitas Semarang, Semarang, Indonesia

ABSTRACT: Companies as an association of stakeholders have a great influence on society and the economy. If the company carries out ethical business practices, the role of community welfare will be greater. This study seeks to find patterns of ethical business practices of public companies in Indonesia. An exploration approach was chosen to extract information from 120 public companies in Indonesia regarding ethical business practices that were carried out. The results of the studies show that ethical business practices are implemented by being integrated into the implementation of good corporate governance and corporate social responsibility to create a sustainable competitive advantage. The findings of this pattern of business best practices in corporations are very helpful for the development of business ethics.

Keywords: stakeholder, practice, business, ethical, corporate governance

1 INTRODUCTION

Ethical business practice is a business that is done in a manner that is ethical or doing business the right way and the best for our stakeholders. A company practicing ethical business practices has advantages over competitors of the company. The ethical business best practice is an advantage that competitors cannot imitate so that the advantage lasts in the long run. Two decades ago, the ethical business was seen as a philosophy only, but now ethical business has become the choice of business strategy (Van Liedekerke and Demuijnck, 2011). Lately, the issue of ethical business has become one of the main topics of research. The company which runs ethical business practices makes business ethics are not just rules that bind the employees and managers in the company, but business ethics into an energy that is able to create added value for the organization. Ethical business practices should have been implemented not only in compliance with laws and regulations but should have chosen as strategies to gain long-term competitive advantage (Crane *et al.*, 2018). Companies that carry out ethical business practices make moral and ethical behavior the core of management strength and the core of business strength.

Ethical business practices in implementation are inherent in the good governance mechanism and implementation of corporate social responsibility. Corporate social responsibility is one of the good governance mechanisms that must be carried out by the company in order to establish a harmonious relationship between stakeholders because the continuity of a business is not only determined by the level of profit, but also corporate social responsibility. The success of a company from the standpoint of corporate social responsibility is to put forward moral and ethical principles, namely, to achieve the best results, without prejudice to other community groups. Companies that work by promoting moral and ethical principles will provide the greatest benefits to society. Not only can the local community be an international community. The implementation of corporate-level corporate social responsibility has been able to contribute to the handling of poverty, inequality, sanitation, etc. (Kolk, 2016).

Corporate social responsibility is rooted in ethics and principles that apply in the Company and in the community. Ethics embraced are part of the culture and corporate cultures are part of the culture of society. The principles or principles that apply in the community also include various government regulations and regulations as part of the state administration system (Baquilas, 2018). Stakeholder theory recommends that companies do a lot of their social responsibility activities because corporate social responsibility brings together the company and its stakeholders. Corporate social responsibility activities enable companies to be closer to customers, local communities or even competitors so that harmonious collaboration is formed. Organizations that can build collaboration with their partners will be able to build a reputation for the company's partners. Collaboration can also reduce the negative effects of information asymmetry. A company is capable of performing a harmonious partnership with stakeholders will have a high reputation recognition for the company's reputation is a multidimensional construct composed of many parties such as customers, investors, employees, and the general public.

2 RESEARCH METHODS

This research is an exploratory study to find patterns of ethical business practices carried out by corporations in Indonesia. The survey was conducted on 120 public companies in Indonesia. Using simple random sampling by sending a questionnaire via e-mail addressed to the Corporate Social Responsibility manager or Corporate Secretary of a public company listed on the Indonesia Stock Exchange.

3 RESULTS AND DISCUSSION

Analysis of respondents' answers regarding ethical business practices carried out shows that companies that carry out ethical business practices are committed to upholding good corporate governance. A series of codes of ethics were agreed upon as a basis for behavior for leaders and employees. The code of ethics forms a work culture that upholds the integrity and harmonious cooperation between stakeholders to create shared prosperity. The results of this exploratory study found that ethical business practices have been carried out by public companies in a formal and systematic manner. Carrying out social responsibility for all stakeholders at the local, national and international levels is also one of the manifestations of ethical business practices.

Ethical business practice is carried out through actions and behaviors that safeguard the business and its employees for professional and legally and morally responsible. Compensation of companies that conduct ethical business practices will carry out programs to guarantee poor employees. In ethical business practices, consumers get the best service and greatest satisfaction. The company that runs public awareness of ethical business practices has urgency in applying the principles of corporate social responsibility wholeheartedly. Corporate Social Responsibility is believed to be a business strategy inherent in the company to maintain or enhance competitiveness through the reputation and image of the company. The implementation of ethical business practices in public companies is largely determined by the leadership and political will of the company leaders and the orientation of shareholders. The harmony of owners and shareholders is the key to implementing ethical business practices that will have an impact on competitive advantage and long-term corporate performance.

Table 1. Description of ethical corporate business practices in Indonesia.

Indicator	Ethical Best Business Practices conducted
The company always strives for the most efficient and ethical way or procedure in all activities	The company implements IT Governance, so that all activities can be systematically documented and can avoid moral hazard. The company provides fast and accurate services with a specific target time The company carries out the production process, safe and energy efficiently.
The company always prioritizes customer satisfaction	The company has the principle of providing the best service and maximum satisfaction to customers. The company provides hotlines, emails, drop boxes and periodical y surveys customer satisfaction. The company provides a whistle blowing mechanism through a Letter to CEO program that can accommodate complaints.
The company always prioritizes professionalism in dealing with each stakeholder	The company upholds and complies with applicable laws and regulations. The company places stakeholders as business partners and parts of the company. The company prohibits vendors from giving anything in kind to members of the company The company establishes detailed and clear code of conduct as a guideline for company members to behave professionally
Every member of the company carries out activities with responsibility and ethics	The company cultivates company members to maintain the company's image The company cultivates full moral and ethical behavior through code of conduct All leaders, both the board of commissioners, CEOs and management ranks create a conducive atmosphere so that members of the company act honestly, keep their promises and uphold values and norms that are in harmony with the principles of good corporate governance in an effort to realize ethical and dignified business. The company encourages company members to be responsible and ethical by holding annual assessments of performance in which there is attitude and ethics Every member of the company is required to fill out and sign annual disclosure, which means that they commit themselves to implementing a code of ethics, oath of office and other applicable regulations. The company is firm in not accepting or giving bribes or gratuities of any kind to company partners The company upholds and maintains fair competition The leaders of all work units are obliged to be role models for employees
The company carries out social responsibility to stakeholders	The company guarantees the welfare of their families and families Companies participate in inclusive community development The company builds synergies with the stakeholders to play a role in community empowerment and concern for the environment. The company supports the achievement of the SDGs goals

Source: primary data

4 CONCLUSION

Ethical business practices, corporate social responsibility, and competitive advantage are three elements that are interrelated with each other. Companies that have a high commitment to run a business based on morals and ethics will implement their social responsibility to create harmony between stakeholders. Harmonious interactions between the company and its stakeholders form a good reputation and corporate image will be a competitive advantage of the company that is difficult to imitate competitors.

REFERENCES

Baquilas, C. J. (2018) 'Training Ethical Culture in the Organization: A Theory on the Role of Ethical Leadership in Humanistic Organizational Culture', *Presented at the DLSU Research Congress 2018De La Salle University, Manila, Philippines.*

Crane, A. *et al.* (2018) 'Business Ethics. Managing Corporate Citizenship and Sustainability in the Age of Globalization', *Oxford University Press*, 1, p. 9.

Kolk, A. (2016) 'The social responsibility of international business: From ethics and the environment to CSR and sustainable development', *Journal of World Business*. Elsevier, 51(1), pp. 23–34.

Van Liedekerke, L. and Demuijnck, G. (2011) 'Business ethics as a field of training, teaching and research in Europe', *Journal of Business Ethics*. Springer, 104(1), pp. 29–41.

Facing Global Digital Revolution – Nirmala Arum Janie, Dwi Mulyaningsih & Wahyu Rachmawati (eds)
© 2020 Taylor & Francis Group, London, ISBN 978-0-367-33912-8

Factors affecting good governance of performance management

B. Hutahayan
Brawijaya University, Malang, Indonesia

ABSTRACT: The aim of this research was to examine the effects of good governance on performance management. This research used a combined quantitative and qualitative approach. The results showed that good governance has positive significant effects on performance management. This study focused on Slank Management, owned by Slank-MANIFESPLUR.

1 INTRODUCTION

Accounting research in the music industry is still very rare and almost untouched by academics. The Indonesian music industry is the largest in Southeast Asia (KS, 2013). This research aims to expand study in the field of good governance to various types of organizations, including the music industry in Indonesia.

Musicians who play together professionally will automatically face economic relations with each other (Garon, 2015). Everyone involved has a relation of partnership. This relation of partnership has two attributes, namely profit sharing and joint control (Okorocha, 2011). Musicians get paid for every concert and for their share of album sales. Joint control includes activities such as choosing when and where to carry out live performances and deciding whether to sign with other parties. Finally, this partnership can lead to conflict, and so good governance becomes very important in its implementation.

Each industry has different business management. Revenue in the music industry is earned through live performances, music sales, merchandise royalties, and copyright (Schultz, 2009). The music industry is so complex that it requires a different perspective from other industries to understand it (Oliver, 2010). Musicians tend to prioritize their freedom, and some achieve success or sustainability within the music industry, while others fail.

Based on this description, this research aimed to examine the factors affecting good governance of performance management within the music industry. In this study, the actors involved in Slank Management were used as informants because Slank Management is one of the most successful managers of musicians in Indonesia. The contributions of this research are: (1) applying accounting science, especially in the field of good governance, to the creative industry; (2) studying good governance practices for effective management of musicians in Indonesia by surveying the actors involved in Slank Management.

2 LITERATURE REVIEW

2.1 *Performance management*

Performance management cannot escape the consequences of the identity crisis confronting the field of which it is a part – that is the field loosely referred to as "administration" or "management." A definition that promises to get us quickly out of the quagmire and take us close to defining *performance management* is that provided by Vieg: "the systematic ordering of affairs and the calculated use of resources aimed at making those things happen which we want to happen and simultaneously preventing the developments that fail to square with our intentions" (Vieg).

The upshot of the preceding analysis is that performance management must go hand in hand with conflict anticipation and management. While focusing on pure managerial variables (i.e., the environmental, organizational, and individual variables), particular attention should be given to the impact of political and corporate governance practices and of the prevailing rules regime on conflict and, therefore, on performance.

2.2 *Good governance*

The OECD defines *corporate governance* as follows: "corporate governance is the system by which business corporations are directed and controlled. The corporate governance structure specifies the distribution of rights and responsibilities among different participants in the corporation, such as the board, the managers, shareholders and other stakeholders and spells out the rules and procedures for making decisions on corporate affairs" (OECD, 1999: 30).

The OECD also states that good corporate governance is a means of managing a company (the directors) that is responsible to the owner of the company or its shareholders. The goal of good corporate governance, as stated in the OECD, is: (1) to reduce the gap between the parties that have an interest in a company; (2) to increase investors' confidence; (3) to reduce the cost of capital; (4) to assure all parties of legal commitments in the management of a company; and (5) to create value for a company, including its relationship with stakeholders (OECD, 1999: 34).

2.3 *Business model in the music industry*

Management of musicians varies, from individuals to limited companies (PTs). Nevertheless, even for the management of musicians with PTs, the good governance model formulated in the foregoing description cannot match fully with the characteristics of a management organization in the music industry. So the model of good governance that is applied must adapt to the music industry. The music industry consists of two components: the concert industry and the recording industry (Koster, 2008).

The musical business model defined generally by Bourreau, Gensollen, and Moreau (2008) is as follows: (1) hit and run, which involves record companies selling traditionally and consumers buying their products; (2) jingles, which means musicians or companies earn income from making jingles; (3) the happy few, which describes a favorable strategy for some consumer segments; (4) net label, which explains how musicians can promote and sell their own music; and (5) "consumartist," which means consumers can also be musicians. The net label and the "consumartist" are new models in the music industry that are currently growing and that can be a threat as well as an opportunity for musicians (Dellyana and Simatupang, 2014).

3 RESEARCH METHODS

This study used a mixed research method. The quantitative part, based on the planned research method, comprised explanatory research, which explains a causal relationship between variables (Cooper and Schindler, 2006: 154). Furthermore, to obtain the data needed to prove the hypothesis of the study, multiple data collection techniques were employed, i.e. questionnaires, interviews, and documentation. The population of this study consisted of people involved in music management in Indonesia.

This research used an interpretive qualitative paradigm. An interpretive paradigm focuses on understanding the social world as it is at the level of subjective experience (Burrell and Morgan, 1979: 28). With it the researcher seeks an understanding of the most basic source of social reality. This paradigm is a direct derivative of the German idealist tradition, which emphasizes the role of language, interpretation, and understanding (Mulawarman, 2010).

4 ANALYSIS RESULTS

4.1 *Quantitative results*

The analysis results, which were arrived at by using structural equation modeling based on partial least square (PLS) analysis, are shown in Figure 1.

The results showed that good governance has a significant positive effect on performance management. The better the implementation of good governance, the better the performance management. Good governance was measured by four indicators, with the dominant factor being transparency. The good governance of Slank Management needs transparency as its first aspect, indicating that it is implementing good governance in its decision-making process and in providing relevant information about the company. Transparency relates to the quality of the information that companies convey. Investor or consumer (the audience) confidence will greatly depend on the quality of information given by a company (Slank Management). Performance management was measured by five indicators, with the dominant factor being artist management. In this case, artists were managed through negotiation, contracts, organized and clear schedules, correspondence, stationery, petty cash, transportation, accommodation, consumption, care for artists' personal needs, coordination with a liaison officer/committee on facilities prepared for artists, media interviews, and technical production of performances with promoters.

4.2 *Good governance from the perspective of the board of commissioners and artists*

One member of the board of commissioners of this company also serves as an artist, Bimbim. From the perspective of Commissioner Bimbim, the information coming from the company is very transparent, with the use of email technology in each report. Bimbim feels that management, in financial terms, is always transparent.

Bimbim's statement came from the standpoint of a company-managed artist. Information, including company activities and financial information, has been delivered by email directly so that Slank personnel get the information in a timely, adequate, and accessible manner.

In terms of accountability, Bimbim as a commissioner still exercises control over management. This is very easy for Bimbim to do considering his double role as an artist and commissioner and that the office is located within his own home. So Bimbim also feels that the company has upheld accountability.

In the music industry, independence means being independent in the management of a company (meaning independent of the company's organs), but it also means being independent in attitude, thought, and creative work. The music business can still run well and support professional work with this unique concept of freedom.

4.3 *Good governance from Slank Management's view*

In terms of transparency, Slank Management provides complete information through its website and YouTube account, and it utilizes social media platforms like Instagram on a regular basis. The website presents information that is always updated related to news, Slank artist

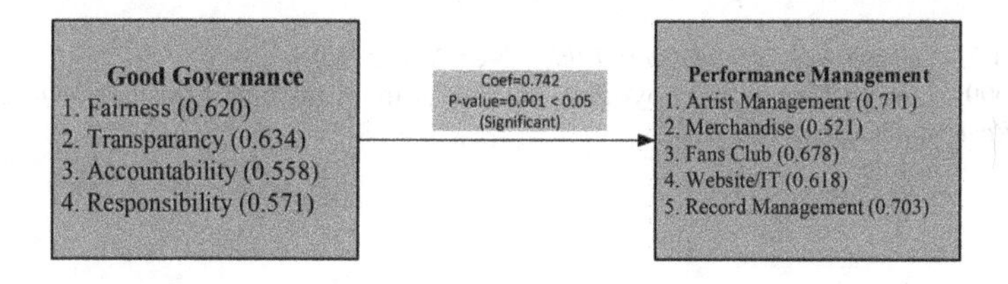

Figure 1. Analysis results.

profiles, discography, the SlankShop, and the upcoming concert schedule, which must be the most material information. Transparency is very easy to provide to "Slankers" with the development of these communication technologies.

In its pursuit of accountability, the company has a section that handles fan clubs that are in charge of processing the SFC's inauguration requirements in branches, handle proposals for proposed SFC activities and report on its activities, maintain an SFC membership database, and encourage and coordinate SFC branch activities. These fan clubs are also required to provide the latest Slank album information. Thus the company also has information on album production needs.

5 CONCLUSION

Conclusions from this study were as follows: (1) The results showed that good governance has a significant positive effect on performance management. The better the implementation of good governance, the better the performance management. Good governance was measured by four indicators, with the dominant factor being transparency. (2) This study finally found the value underlying good governance at Slank Management, owned by Slank-MANIFESPLUR. The fundamental difference lies in the fact that the main stakeholder position for Slank is not shareholders, but Slank's own personnel and Slankers as the main consumers of Slank's products. Thus good governance in Slank Management is directed to manage business in such a way that artistic activity is always supported. The good governance found in Slank Management has a strong foundation.

REFERENCES

Bourreau, M., Gensollen, M., and Moreau, F. (2008). The digitization of the recorded music industry: Impact on business models and scenarios of evolution.

Burrell, G. and Morgan, G. (1979). *Sociological Paradigms and Organisational Analysis*. London: Heinemann.

Chua, W. F. (2008). Radical developments in accounting thought. *Accounting Review*, 61(4), pp. 601–632.

Dellyana, D. and Simatupang, T. M. (2014). Existing music business model in Indonesia in search of new income sources. *Procedia: Social and Behavioral Sciences*, 115, pp. 407–414. http://doi.org/10.1016/j.sbspro.2014.02.447

Garon, J. M. (2015). The band: Artistic, legal, and financial structures which shape modern music. *Music and Law*, pp. 79–99.

Koster, A. (2008). The emerging music business model: Back to the future? *Journal of Business Case Studies*, 4(10), pp. 17–22.

KS, T. (2013). *Rock 'n Roll Industri Musik Indonesia: Dari Analog ke Digital*. Jakarta: Penerbit Buku Kompas.

Mulawarman, A. D. (2010). Integrasi paradigma akuntansi: Refleksi atas pendekatan sosiologi dalam ilmu akuntansi. *Jurnal Akuntansi Multiparadigma*, 1(1), pp. 155–171.

Okorocha, D. (2011). A full 360: How the 360 deal challenges the historical resistance to establishing a fiduciary duty between artist and label. *UCLA Entertainment Law Review*, 18(1), pp. 1–31.

Oliver, P. G. (2010). The DIY artist: Issues of sustainability within local music scenes. *Management Decision*, 48(9), pp. 1422–1432.

Organisation for Economic Co-operation and Development. (1999). *OECD Principles of Corporate Governance*.

Rezaee, Z. (2009). *Corporate Governance and Ethics*. New York: Wiley.

Schultz, M. (2009). Live performance, copyright, and the future of the music business. *University of Richmond Law Review*, 43(2), pp. 685–764.

Facing Global Digital Revolution – Nirmala Arum Janie, Dwi Mulyaningsih & Wahyu Rachmawati (eds)

Dynamic capabilities of product-service system implementation: The Indonesian industrial estate firm evidence

Christina Wirawan
School of Business Management, Bandung Institute of Technology, Bandung, Indonesia
Industrial Engineering Department, Universitas Kristen Maranatha, Bandung, Indonesia

Gatot Yudoko & Yuliani Dwi Lestari
School of Business Management, Bandung Institute of Technology, Bandung, Indonesia

ABSTRACT: Indonesian industrial estate firms play essential roles in their contribution to national economic development. This article discusses how industrial estate firms could be sustainable in each income condition despite limited tangible resources and difficult business situations. The product-service system (PSS) concept aims to help industrial estate firms' sustainability within various conditions by changing the bundle of products and services in order to gain new sources of income. To enable the dynamicity of PSSs, firms need to have this dynamic capability. Thus, this article focuses on three case studies to test this proposition. The novelty of this article is its consideration of the implementation of dynamic PSSs at industrial estate firms. This concept is needed since PSSs deal with various conditions and can be made more sustainable. This article not only offers industrial estate firms best practice for sustainability but also helps the government to create and implement supporting policies.

Keywords: dynamic capability, indonesian industrial estate firms, product-service system, sustainability

1 INTRODUCTION

Industrial estate firms (IEF) comprise one of the essential sectors in Indonesia that provides a significant contribution to national economic development, as well as assisting the government tin environmental and social sustainability (Republic of Indonesia, 2014). Considering the important role of IEFs, it is imperative that they be sustained.

The unique problem IEFs encounter is limited tangible resources (e.g., buildings and land). Their profits will decrease when the sales of buildings and land are reduced due to fewer buildings and less available land. Despite the business situation of industries that significantly affect IEFs' income, they need to keep playing their roles properly.

Considering these issues, in order to be sustained, IEFs need to think of a new business model exploiting services delivery (Mulyadi, 2012). The product-service system (PSS) concept offers an answer to the problem since it tries to shift tangible product into a bundle of tangible products and services(Mont, 2002). Due to differences in the condition of various IEFs, different packages of PSS are essential. In other words, there is variation in the appropriate PSSs. To enable the IEFs to implement the relevant dynamic that PSSs need, firms must develop the capabilities of sensing, seizing, and reconfiguring (Kindström, Kowalkowski, and Sandberg, 2013; Teece, Pisano, and Shuen, 1997).

Dynamic capability and its impact on organizational performance has been broadly researched and discussed by experts (e.g., Kuo, Lin, and Lu, 2017), as have competitive advantage (e.g., Fainshmidt,Wenger, Pezeshkan, and Mallon, 2019), dominant logic change (e.g., Ellonen, Jatunen, and Johansson, 2015), and service innovation (Kindström et al., 2013). However, gaps related to PSS dynamics still exist. This article contributes to extending dynamic capability knowledge related to PSS dynamics enabling at IEFs, and aims to help IEFs to sustain and play their roles properly, as well as to help the government to build suitable policies to support IEFs.

2 LITERATURE REVIEW

In Indonesia, the IEF industry was deemed one of the national economic development pillars and therefore it must operate in industrial estates, except for a few exceptional cases(Republic of Indonesia, 2014). An industrial estate is a location intended for industries operating and equipped with infrastructure, and amenities thus supporting the efficiency, effectiveness, and productivity of enterprises operating in it (Fonseca, 2015). The other roles of IEFs include helping the government in protecting the environment and developing social welfare, as well as attracting foreign investors (Republic of Indonesia, 2014).

Indonesian IEFs need to gain profits in order to be sustainable and play their role of supporting tenants appropriately. The most substantial portion of IEFs' profits comes from selling land and buildings or leasing them for the long term. But land and buildings are limited resources. A time will come when the land and buildings are sold out. The IEFs will only rely on service fee revenues paid by tenants for provided infrastructures; this will not provide sufficient income. Moreover, industrial estates' income significantly depends on national and global business conditions. Therefore, IEFs need to look for new sources of income from service-based innovations as a new business model.

The PSS concept, which offers various bundles of products and services, would be a suitable concept to solve the problem. The PSSs shift products and services into their various bundles(Mont, 2002). Previous researchers have identified a continuum of products and services bundles starting from a pure product, then over time the proportion of the product decreases (Tukker, 2004). This continuum can be applied suitably across the various conditions of IEFs. To create a suitable bundle of products and services, firms need to have dynamic capabilities. They serves as the key to building firms' resources and competencies by adapting to changes in the business environment (Adner and Helfat, 2003).

Many definitions and understandings of dynamic capabilities have existed. However, the most used definition is the ability of firms to overcome a changing environment through changes their set of resources and methods (Teece et al., 1997).Dynamic capability includes sensing threats, seizing opportunities, and reconfiguring resources (Teece et al., 1997).

The dynamic capability concept has been well researched already in different aspects and with various focuses (e.g., Kuo et al., 2017; Schilke, Hu, and Helfat, 2018). Nevertheless, gaps within opportunities for further research still exist (Schilke et al., 2018). We catch one of the research opportunities, to empirically investigate a concept (PSS) using a dynamic capabilities perspective, as suggested by Schilke et al. (2018). We also extend the work of Kindström et al. (2013) to identify sensing, seizing, and reconfiguring practices in IEFs.

3 METHODOLOGY

We used the case study method to study the interactions of actors. This provided chances to acknowledge as well as reveal the evidence, which is appropriate when building a conceptual framework. We took three steps: (1) concept identification, (2) concept organization, and (3) boundaries determination (Swanson and Chermack, 2013).

The first step was concept identification. Concepts used in this article include dynamic capability and the product-service system. The second step was concept organization. In this article, dynamic capability is positioned as an enabler of the PSS dynamic, coping with the up-and-down situation of IEFs. To study and confirm the conceptual framework, we conducted interviews with three privately owned IEFs that have operated for more than 20 years, as proof of their sustainability. The interviewees were the managers, division heads, and staff who were involved in the development of the IEFs. Besides the interview, we also collected various data from documents and websites. Additional information was also obtained from interviews with the executive director of the IEFs' association (HKI) and with industrial regional development staff at the Indonesian Industrial Ministry.

The last step was boundaries determination. Because the data were collected from Indonesian IEFs, the conclusions drawn were for Indonesian IEFs.

4 RESULT AND DISCUSSION

According to the data collected, most Indonesian IEFs have been aware of the need to explore service innovation to generate new income when the land and buildings run out. They tend to search for a new business model that is mostly based on services delivery. In other words, they already implement PSS in their various bundles of products and services (Wirawan, Yudoko, and Lestari, 2017).

In this article, we discuss three privately owned IEFs, namely IEF 1, IEF 2, and IEF 3. These industrial estates had already operated for more than 20 years at the time of our research and are still well operating. IEF 1, located in Java, had all of its land and buildings sold out. It shifted its business to providing various services, such as landscape gardening, health care facilities, drinking water, and other service innovations. IEF 2, also located in Java, still own limited land and buildings. But it realized the importance of service innovation as a new source of income to sustain itself. IEF 2 builds commercial areas, warehouses, and ready-to-use set factories for rent. IEF 3 is located outside Java Island. Currently, IEF 3 still owns land and buildings. However, since 2012, IEF 3 has merely rented land and buildings rather than selling them. IEF 3 learned from IEFs that declined after their land and buildings ran out. Innovations developed by IEF 3 seek efficiency, effectiveness, and quality in environmental management.

These three IEFs sensed threats and found different opportunities in the form of varying PSSs. Then, they seized the opportunities by cultivating their resources accordingly. In the end, they all reconfigured to adapt their resources with the opportunity that existed. Table 1 presents details of the dynamic capability component these three IEFs practices in creating various bundles of PSSs for each condition.

Table 1. Dynamic capability practices of IEF 1, IEF 2, and IEF 3.

	IEF 1	IEF 2	IEF 3
Sensing	• Tenants' needs • Environmental protection needs • Regulations that create opportunity	• Tenants' needs • Environmental protection needs • Regulations that create opportunity	• Tenants' needs • Environmental protection needs • Regulations that create opportunity • Other IEFs' downturn experience
Seizing	• Install waste and water treatment before sell land and buildings to tenants and maintain the best waste and water treatment • River as the water source • Excellent services for tenants • Identify resource needs to built services needed by tenants and catch regulation	• Excellent services for tenants • Space to built warehouse and factory • Commercial area • The capability to set factory for rent • The capacity to meet the environmental requirement • Having a culture tries to do the best to meet the needs of tenants	• A favourable location close to Singapore • Quality circle convention with tenants • A policy to process environmental aspect to gain effectiveness and efficiency that lead to environmental protection as well as a cost saving • Unsold land and building
Recon-figuring	• Using the existing and create the required resources	• Using the existing and create the required resources	• Using the existing and create the required resources

5 CONCLUSIONS

We found that each PSS type has different capabilities that require various resources (Wirawan et al., 2017). Given this fact, dynamic capabilities are suitable to enable the PSS dynamic.

The three privately owned IEFs examined here already have the dynamic capability to cope with the changing situations that IEFs face. The sensing and seizing results are different according to their opportunity and resources, but they all follow a similar procedure. With these data, it can be shown that dynamic capability enables the PSS dynamic to sustain IEFs.

This article only discussed three privately owned IEFs. Further research can be conducted by considering more IEFs, comparing privately owned to state-owned IEFs, and comparing IEFs inside Java and outside Java. Further research also can be conducted by studying the dominant logic change underlying PSS dynamics.

This article brings a scientific contribution to shed light about PSSs. Moreover, there are managerial implications to help IEFs to sustain themselves in changing conditions as well as to help the government built a policy to support IEFs' sustainability.

ACKNOWLEDGMENT

The authors thank the Indonesian Endowment Fund for Education (LPDP) and Higher Education for funding this research, and Ms. Prameshwara Anggahegari for proofreading this article.

REFERENCES

Adner, R. and Helfat, C. E. (2003).Corporate effects and dynamic managerial capabilities. *Strategic Management Journal*, 24(10), pp. 1011–1025.

Ellonen, H.-K., Jatunen, A., and Johansson, A. (2015).The interplay of dominant logic and dynamic capabilities in innovation activities. *International Journal of Innovation Management*, 19 (5),pp. 1550032-1–1550052-15.

Fainshmidt, S., Wenger, L., Pezeshkan, A., and Mallon, M. R. (2019).When do dynamic capabilities lead to competitive advantage? The importance of strategic fit. *Journal of Management Studies*, 56(4), pp. 758–787.

Fonseca, F. (2015). An agent based model to assess the attractiveness of industrial estates. *Journal of Artificial Societies and Social Simulation*, 18(4), pp. 1–11.

Kindström, D., Kowalkowski, C., and Sandberg, E. (2013). Enabling service innovation: A dynamic capabilities approach. *Journal of Business Research*, 66(8), pp. 1063–1073.

Kuo, S., Lin, P., and Lu, C. (2017). The effects of dynamic capabilities, service capabilities, competitive advantage, and organizational performance in container shipping. *Transportation Research Part A*, 95, pp. 356–371.

Mont, O. K. (2002).Clarifying the concept of product-service system. *Journal of Cleaner Production*, 10 (3), pp. 237–245.

Mulyadi, D. (2012). *Manajemen Perwilayahan Industri*. First edition. Jakarta: Leuser Cita Pustaka.

President of the Republic of Indonesia (1989). Keputusan Presiden Republik Indonesia No. 53/1989 Tentang Kawasan Industri.

Schilke, O., Hu, S., and Helfat, C. E. (2018). Quo vadis, dynamic capabilities? A content-analytic review of the current state of knowledge and recommendations for future research. *Academy of Management Annals*, 12(1), pp. 390–439.

Swanson, R. A. and Chermack, T. J. (2013). *Theory Building in Applied Disciplines*. San Francisco: Berret-Koehler.

Teece, D. J., Pisano, G. and Shuen, A. (1997). Dynamic capabilites and strategic management. *Strategic Management Journal*, 18(7), pp. 509–533.

Republic of Indonesia (2014). Undang-Undang Republik Indonesia Nomor 3 Tahun 2014 Tentang Perindustrian (The Republic of Indonesia Law no. 3 the year of 2014).

Tukker, A. (2004). Eight types of product-service system: Eight ways to sustainability? Experiences from suspronet. *Business Strategy and the Environment*, 13(4), pp.246–260.

Wilden, R., Gudergan, S., Akaka, M. A., Averdung, A., and Teichert, T. (2019). The role of co-creation and dynamic capabilities in service provision and performance: A configurational study. *Industrial Marketing Management*, 78, pp. 43–57.
Wirawan, C., Yudoko, G., and Lestari, Y. D. (2017). Product-service system for Indonesian industrial estate firms: A conceptual framework. In *Proceedings of 2017 IEEM*. Singapore: IEEE, pp. 1812–1816.

Facing Global Digital Revolution – Nirmala Arum Janie, Dwi Mulyaningsih & Wahyu Rachmawati (eds)
© 2020 Taylor & Francis Group, London, ISBN 978-0-367-33912-8

Author Index

Author Index